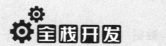

深入浅出
React Native

陈陆扬 / 著

人民邮电出版社

北京

图书在版编目（CIP）数据

深入浅出React Native / 陈陆扬著. -- 北京 : 人民邮电出版社, 2021.12（2023.3重印）
ISBN 978-7-115-57242-4

Ⅰ. ①深… Ⅱ. ①陈… Ⅲ. ①移动终端－应用程序－程序设计 Ⅳ. ①TN929.53

中国版本图书馆CIP数据核字（2021）第178942号

◆ 著　　　陈陆扬
　责任编辑　赵　轩
　责任印制　陈　犇

◆ 人民邮电出版社出版发行　北京市丰台区成寿寺路11号
邮编　100164　电子邮件　315@ptpress.com.cn
网址　https://www.ptpress.com.cn
北京捷迅佳彩印刷有限公司印刷

◆ 开本：800×1000　1/16
印张：17.5　　　　　　　　　2021年12月第1版
字数：396千字　　　　　　　2023年3月北京第2次印刷

定价：69.90元

读者服务热线：（010）81055410　印装质量热线：（010）81055316
反盗版热线：（010）81055315
广告经营许可证：京东市监广登字 20170147 号

前　言

跨平台方案一直是端设备应用开发的热点。一直以来，设计者和开发者都在追求更好的产品体验与降低开发成本、提高响应速度之间寻求平衡。而 React Native 也是诸多方案之一，面世之初，它以基于 React 框架的高效开发方式，和优于 Hybrid Webview 方案的用户体验受到了广大开发者的关注。但随之而来的是人们对其质量、稳定性和性能方面的质疑。甚至在一些场景中，JavaScript 开发者并不能完全依赖 React Native 提供的方式，依旧需要 iOS/Android 工程师的支持，并不能做到完全屏蔽平台差异。

除了技术难点之外，跨端配合也是实际开发中经常遇到的问题。React Native 主要的作用是让 JavaScript 开发者直接开发原生应用，这对原生开发者来说或多或少挤占了一部分开发空间，双方目标不一致也就很难协同。但整体来看，没有一项技术能够脱离产品和业务单独存活，一项能够降低开发成本的技术一定有其生存的空间。对于大多数的开发者而言，技术价值的体现基本都包括在产品的实现以及对业务结构的提升上。但如果新技术在节省某些日常基础开发时间的同时，增加了自己的学习成本，我们又该持什么样的态度去面对这种变化和挑战呢？

本书面向 JavaScript 开发者深度分析了 React Native 实现的原理，包括 iOS/Android 中常见的开发概念及实现方式，帮助 JavaScript 开发者更好地理解原生端开发的特点；也面向 iOS/Android 开发者，解释 React Native 利用了哪些原生特性，两端之间如何配合及通信，并且说明了从操作系统的角度看待自身平台，对于上层的实现能有更好的支持。本书的所有代码均可在 GitHub 上的 react-native-explorer 项目中找到。

最后，感谢杨攀、赵延达、付晓龙、吴汀含 4 位 iOS/Android 工程师的大力支持，书中涉及的原生能力的扩展和探索方案的具体实现，均由大家共同完成，没有不同平台开发者的协作，本书也就不可能完成。

目　录

第1章　走进React Native … 1
1.1　React Native给我们带来了什么 … 1
1.2　React Native的适用场景 … 2
1.3　搭建React Native环境 … 2
1.3.1　iOS开发常见概念 … 2
1.3.2　Android开发常见概念 … 6
1.3.3　命令行构建 … 8
1.3.4　在现有原生项目中增加React Native环境 … 9
1.4　本章小结 … 15

第2章　React Native启动流程及视图解析 … 16
2.1　React Native启动流程 … 16
2.1.1　iOS启动流程 … 17
2.1.2　Android启动流程 … 18
2.1.3　小结 … 20
2.2　局部渲染React Native … 20
2.2.1　iOS局部渲染 … 20
2.2.2　Android局部渲染 … 21
2.3　React Native原生视图详解 … 23
2.3.1　iOS——RCTRootView … 23
2.3.2　Android——ReactRootView … 24
2.3.3　视图长度单位 … 26
2.4　React Native布局方式 … 28
2.4.1　Flex布局 … 28
2.4.2　绝对定位 … 32
2.5　本章小结 … 34

第3章　文本及输入 … 35
3.1　Text解析 … 35
3.1.1　RCTTextView和ReactTextView … 37
3.1.2　行间距 … 39
3.2　Text布局方式 … 41

		3.2.1	Text的嵌套	41
		3.2.2	同行多字号文本的对齐方式	43
	3.3	文本输入——TextInput		45
	3.4	软键盘		49
		3.4.1	Keyboard	49
		3.4.2	KeyboardAvoidingView	51
	3.5	本章小结		52

第4章 事件响应机制 ········· 53

	4.1	触摸事件		53
	4.2	Touch组件		55
	4.3	手势响应系统		58
		4.3.1	响应者生命周期	58
		4.3.2	PanResponder	64
	4.4	原生事件机制		67
		4.4.1	iOS事件机制	68
		4.4.2	Android事件机制	70
	4.5	本章小结		74

第5章 媒体、文件及本地存储 ········· 75

	5.1	图片组件		75
		5.1.1	Image属性及方法详解	77
		5.1.2	原生图片组件	81
		5.1.3	高性能图片组件：react-native-fast-image	83
	5.2	音视频文件的操作方式		86
		5.2.1	音频处理	86
		5.2.2	视频处理	94
	5.3	本地文件系统		98
		5.3.1	iOS本地文件系统	98
		5.3.2	Android本地文件系统	99
		5.3.3	react-native-fs	100
	5.4	本地存储		103
		5.4.1	iOS本地存储方式	103
		5.4.2	Android本地存储方式	104
		5.4.3	React Native本地存储方式	108
		5.4.4	React Native混合模式下的公共存储方案	110

5.5　本章小结 ··116

第6章　动画 ···117

6.1　布局动画——LayoutAnimation ··117
　　6.1.1　基本用法 ···117
　　6.1.2　原生实现原理 ···124
6.2　交互动画——Animated ···128
　　6.2.1　基本用法 ···128
　　6.2.2　动画的控制与组合 ···139
　　6.2.3　动画值的运算与变化 ··143
　　6.2.4　手势跟踪 ···145
6.3　动画实现原理及优化 ···146
　　6.3.1　动画实现原理 ···147
　　6.3.2　常见优化手段 ···148
6.4　本章小结 ··152

第7章　React Native与原生端的通信方式 ···153

7.1　JavaScript调用原生模块 ··153
　　7.1.1　iOS与JavaScript的通信方式 ··153
　　7.1.2　Android与JavaScript的通信方式 ···160
7.2　JavaScript跨平台运行原理 ···169
　　7.2.1　JavaScriptCore——iOS的JavaScript引擎 ·································169
　　7.2.2　Hermes——Android的新版JavaScript引擎 ·······························173
7.3　本章小结 ··177

第8章　自定义原生组件 ···178

8.1　原生UI组件封装 ···178
　　8.1.1　iOS原生组件封装 ··178
　　8.1.2　Android原生组件封装 ···181
　　8.1.3　JavaScript直接调用原生组件方法 ···186
8.2　自定义插件 ··189
8.3　本章小结 ··190

第9章　React Native的导航方案 ··191

9.1　原生导航偏好 ···191
9.2　JavaScript导航——React Navigation ···192
　　9.2.1　自定义导航 ··193

9.2.2 导航事件 ·············· 198
9.3 原生导航——React Native Navigation ·············· 199
9.3.1 自定义导航 ·············· 200
9.3.2 视图生命周期 ·············· 204
9.4 混合导航探索 ·············· 205
9.4.1 方案设计 ·············· 206
9.4.2 扩展功能 ·············· 214
9.5 本章小结 ·············· 218

第10章 热更新与多实例 ·············· 219
10.1 热更新 ·············· 219
10.1.1 热更新流程 ·············· 219
10.1.2 第三方服务 ·············· 220
10.1.3 具体实现 ·············· 221
10.2 App平台化——React Native多实例 ·············· 225
10.2.1 多实例管理 ·············· 225
10.2.2 指定渲染依赖实例 ·············· 226
10.2.3 自定义原生模块依赖 ·············· 230
10.2.4 多实例效果及局限 ·············· 233
10.3 本章小结 ·············· 235

第11章 常见场景优化 ·············· 236
11.1 页面启动白屏时间 ·············· 236
11.1.1 JavaScript Bundle包大小的影响 ·············· 236
11.1.2 自定义原生模块的影响 ·············· 237
11.1.3 页面层级深度 ·············· 238
11.2 视图预加载 ·············· 241
11.3 长列表优化 ·············· 250
11.3.1 FlatList、SectionList和VirtualizedList ·············· 251
11.3.2 原生视图的复用 ·············· 254
11.4 Tab切换 ·············· 263
11.5 本章小结 ·············· 264

第12章 React Native中的"微前端" ·············· 265
12.1 什么是微前端 ·············· 265
12.2 React Native"微前端"探索 ·············· 271
12.3 本章小结 ·············· 272

第 1 章　走进 React Native

自 2013 年 Facebook 团队（后文简称为 Facebook）发布 React 框架后，这种新型的 Web 开发技术得到了广泛的应用和支持，极大程度上提升了 Web 开发的效率，并且降低了开发复杂 Web App（应用或应用程序）的难度。不过，Web 应用毕竟只是互联网开发的一部分，随着移动互联网的发展，越来越多的 App 开发需求如雨后春笋般冒出，对研发的效率也提出了更高的挑战。那么在 Web 开发中大放异彩的 React 框架，是否能够应用到移动端开发中，从而也提升移动应用的开发效率呢？2015 年，Facebook 发布了 React Native。React Native 利用 React 组件化及虚拟 DOM 的特性，建立了 JavaScript 到原生视图的映射关系，可以将 JavaScript 开发者编写的 React 工程转化成包含 iOS 及 Android 的原生应用，并尽可能保留原生应用的体验，同时降低了开发成本。

由于目前 React Native 仍在不断升级版本，为确保一致性，本书采用的是 0.60 版本，下文所介绍的 React Native 均默认为该版本。

1.1　React Native 给我们带来了什么

在 React Native 诞生之前，为了节省成本，开发者在开发 App 时会选择用 WebView 封装 H5。如果项目需要额外原生能力的话，则可以采用 Cordova(PhoneGap) 这类框架和工具，也可以达到一定的效果，但这种混合方式始终在体验和性能方面与原生应用存在着一定的差距。而使用 React Native 开发的项目，其最终生成的代码及项目仍是一个原生应用，只不过将页面渲染、事件交互等工作交给了 JavaScript 端处理，与纯粹的原生应用相比多了一层与 JavaScript 端通信的成本。但与使用 WebView 相比较，使用 React Native 开发的应用在体验和性能上已经有了很大的提升，并且再也不会受到 WebView 的限制。

那么，React Native 具体给我们提供了哪些开发方式和现成的工具呢？

（1）使用 React 的方式开发原生应用，包括完整的构建及打包机制。

（2）超过 20 个基础 UI 组件，包括基础视图、文字、图片、输入框等，并且支持 Flex 布局。

（3）手势事件响应体系，用于处理用户交互。

（4）可调用的原生API，包括设备属性、动画和本地存储等。

（5）丰富的社区资源和插件。

其中，React Native官方资源列表（官网more-resources目录下的Awesome React Native）里面罗列了社区生态中关于React Native的资料、工具和组件等一系列开源项目，方便开发者选择适用于当前自身需求的方案。

1.2 React Native的适用场景

那么，在什么场景中适合采用React Native方案呢？这估计是被讨论次数最多，且永远不会有结论的问题；或者说最后的答案始终会是"根据实际情况"。随着2018年6月Airbnb宣布放弃使用React Native，就有不少言论不看好React Native，大家纷纷觉得应该回归原生开发或者探求新的高性价比开发方式。或许我们始终不会找到能够覆盖所有范围、适用所有场景的跨平台技术方案，但对这些技术方案了解得越多、越深入，你就越需要根据实际场景来选择技术方案。

通常而言，以下几种场景更适合使用React Native开发方案。

（1）需求变动频繁：原生开发流程较长，但诸如营销类的需求则需要对节庆、热点进行快速响应，并且流程环节可能会经常发生变化，React Native的热更新、跨平台机制则可满足这种需求。

（2）内部OA：越来越多公司的OA系统有着移动化的趋势，对于一些交互简单但流程较长的工作场景，例如审批、请假，大家会更偏向于在移动端处理。这类系统对体验的要求相对低一点，使用者又大部分是内部员工，使用React Native开发方案则能极大地降低开发成本。

1.3 搭建React Native环境

搭建环境通常是学习一门新技术时最先遇到的难题，特别是在跨平台开发中，对刚入门的开发者来说是不小的挑战。在搭建环境之前，我们先大致解释一下各平台上的常见概念及其作用，以减少学习者的陌生感。

1.3.1 iOS开发常见概念

iOS提供了从开发语言、IDE到App发布及付费的闭合生态，其中包含了大量平台独有的概念及开发步骤。

1. Xcode

Xcode是开发iOS App的IDE，用于开发、打包及发布iOS App。你可以通过App Store下载最新的Release版本，也可以通过Apple Develop下载beta版。需要注意的是，beta版的Xcode无法提交构建好的App到iTunes。

2. CocoaPods

CocoaPods是开发者必备的iOS依赖管理工具。在安装CocoaPods之前，首先要在本地安装好Ruby环境，之后只需要执行sudo gem install cocoapods命令就可以直接安装了。

3. 模拟器（Simulator）

使用模拟器（Simulator）可以在mac OS上模拟iPhone、iPad。下载Xcode时，最新的模拟器会自动一起安装到本地。如果需要增加机型，可以选择addDevice自行添加。如果需要使用旧版本的iOS，则需要在Xcode > Preferences > Components中自行下载，之后所有支持该系统版本的机型都可以使用该系统。

4. 打包发布

关于iOS App打包发布需要了解两个概念——开发者账号和开发者证书。如果你想把自己的App上传到App Store，需要申请个人开发者账号或企业开发者账号，大致步骤如下。

（1）开发者账号注册：使用Apple ID登录Developer网站，在页面中找到"Join the Apple Developer Program"并按照提示付费和申请，如图1.1所示。

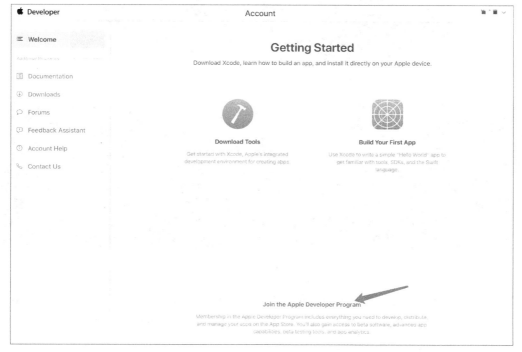

图1.1　Apple开发者首页

（2）证书配置：如果持有开发者账号，可以选择Xcode > Preferences > Accounts，在Accounts界面中直接登录该账号，如图1.2所示，并且在TARGETS选项的Signing & Capabilities标签下勾选

Automatically manage signing（自动管理证书，见图1.3），Xcode会自动使用TARGETS选项中配置的Bundle Identifier去申请本机可使用证书。如果未持有开发者账号，则需要在Apple Develop中创建App并在配置证书的计算机上导出p12文件以及描述文件，安装到当前非开发者账号的设备上，并且在TARGETS选项的Signing & Capabilities标签下选择该证书和描述文件。

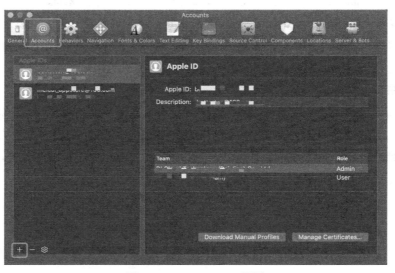

图1.2　Xcode Accounts 配置

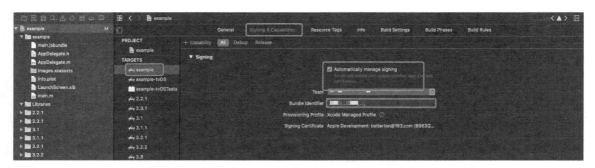

图1.3　Xcode Signing & Capabilities

（3）打包：将device选择到非模拟器条目上（真机或空都可以），选择Product > Archive命令进行打包编译，如图1.4所示。打包完成后会自动弹出Organizer（也可以在Window菜单中找到打开该功能的命令）；选中刚才打包成功的Archive并单击右侧的Distribute App按钮，按照提示选择打包类型，如图1.5所示。如需上传App Store，则可以选择App Store Connect，直接打包成ipa文件并上传到iTunes。ADHoc和企业证书的打包选项，会影响可安装设备以及证书的选择。

图 1.4 打包编译

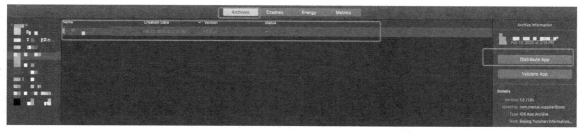

图 1.5 发布

iOS的证书体系相对繁琐，在这里我们简单介绍申请过程中会遇到的一些概念。

Certificates：iOS证书是用来证明iOS App内容合法性和完整性的数字证书。想安装到真机或发布到App Store的App必须经过签名验证，并且其内容是完整、未经篡改的。

Identifiers：App ID，用于标识一个或者一组App，App ID和Xcode中的Bundle ID是一致或匹配的。

Devices：iOS设备。Devices中包含了该账户中所有可用于开发和测试的设备，每台设备使用UDID（设备唯一识别符）来唯一标识。

Profiles：也称Provisioning Profile。一个Provisioning Profile文件包含了上述的所有内容，如证书、App ID、设备，是一个综合描述文件。如果我们要打包或者在真机上运行一个App，首先需要证书来进行签名，用来标识这个App的合法性；其次，需要指明它的App ID，并且验证Bundle ID是否与其一致；最后，在真机调试环节，确认这台设备是否能够运行该App。我们只需选择对应的profile文件就可以进行正常的打包和真机调试，并且这个Provisioning Profile文件会在打包时嵌入.ipa。

上述相关文件均可在Apple开发者网站进行管理，如图1.6和图1.7所示。

图1.6 证书管理入口

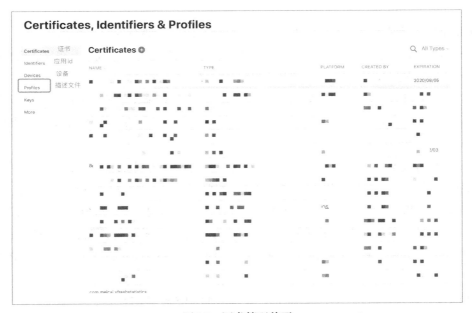

图1.7 证书管理首页

1.3.2　Android开发常见概念

Android是基于Linux的开源操作系统，也是当今最主流的智能手机系统之一。

1. Android Studio

Android Studio集成了Android开发所需要的各种工具，并为Android项目优化了目录文件结构的展示。如果开发者之前没有配置过Android开发环境，建议下载集成了Android SDK的Android Studio安装包，可以极大地减少第一次打开Android Studio时下载相关开发工具所需的时间。

2. Gradle

Gradle是一款支持依赖管理、自动化构建、打包、发布和部署的通用性构建工具，主要使用Groovy语言。开发者可以通过应用不同的Gradle插件来构建不同的工程项目。对Android项目来说，用户必须要在工程根目录的build.gradle文件中首先指明适用于Android项目的构建插件——Android Plugin for Gradle，然后添加配置项，所以在Android工程的build.gradle中经常会看见以下代码。

```
apply plugin: 'com.android.application' // 指明适用于Android项目的构建插件

android { // 为Android添加配置项
    ......
}
```

Android视图是Android Studio的默认文件目录视图模式，如图1.8所示，通过聚合Gradle相关文件，可以减少开发者花费在探索目录上的时间。

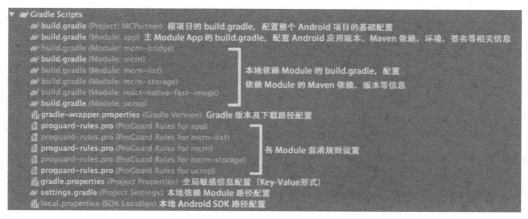

图1.8　Gradle相关文件的作用

3. 模拟器

如果开发Android App时没有真机设备，可以通过Android Virtual Device Manager（AVD Manager）添加模拟器，如图1.9所示。

添加不同配置和规格的模拟器需要再次下载相关文件，可能需要一些时间。由于许多Android机型采用了经过各厂商二次开发的深度定制系统，因此最好使用真机进行开发调试，避免出现兼容性问题。

图1.9　Android Studio模拟器位置

4. 打包

使用Android Studio打包项目生成APK（App包），需要先在左边的Build Variants选项中选择打包环境和debug/release版本，然后单击工具栏中的Run按钮，即可将安装包安装至连接的真机或模拟器。打包生成的APK在/app/build/outputs/apk下，如图1.10所示。

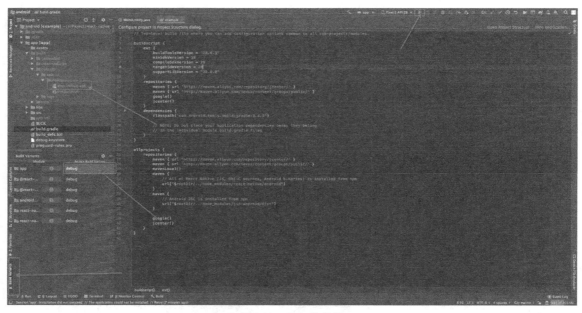

图1.10　APK的位置

1.3.3　命令行构建

和大部分框架一样，React Native也具有自己的命令行工具react-native-cli，用于初始化项目。开发者可以通过npm或yarn命令来安装react-native-cli。

```
npm install -g react-native-cli
```

安装好之后，可以直接通过命令来构建React Native项目。

`react-native init FirstRN`

也可以指定对应的版本号。

`react-native init FirstRN --version 0.60.5 // 注意版本号需要精确到补丁版本号`

执行完成后，你将收获一个最基础的React Native项目，包含了JavaScript所需的依赖以及iOS工程和Android工程。虽然官方文档上说直接进入项目文件夹运行react-native run-ios即可，但根据自身环境不同，所装依赖的版本也有可能不同，会有一定概率无法成功。建议直接使用Xcode和Android Studio来运行项目，里面有比较详尽的报错信息，无论是依赖缺失还是版本不一致，都可以打印出来。

如果一切顺利的话，你应该能看到0.60版本默认页面如图1.11所示，并且在Learn More中包含了一些基本的文档入口，直接单击即可从浏览器打开对应文档，方便查阅。而该页面局部的代码即包含在项目的App.js中，你可以对其进行修改，开发自己的"Hello World"项目。

图1.11　React Native启动页

1.3.4　在现有原生项目中增加React Native环境

利用命令行工具生成工程无疑是最方便快捷的方式，特别是当你需要从零开始搭建项目的时候。但这并不能满足所有的开发场景，有时候可能需要在现有的原生工程中添加React Native环境，将一部分功能交由JavaScript开发。此时，我们要手动在iOS和Android项目中搭建React Native环境。

无论是iOS还是Android项目，在手动搭建React Native环境时都需要通过package.json安装JavaScript依赖。

```
{
  "name": "MyReactNativeApp",
  "version": "0.0.1",
  "private": true,
  "scripts": {
    // react服务启动的脚本
    "start": "yarn react-native start"
  },
  "dependencies": {
    // 可以指定版本号
    "react": "16.8.6",
```

```
        "react-native": "0.60.5",
    }
}
```

然后执行 npm install 下载所需的依赖包。

1. iOS

安装完 JavaScript 依赖后,需要在 podfile 中添加 React Native 依赖,其中的 path 需要指定到上述 package.json 安装的 node_modules 路径下。

```
pod 'React', :path => '../node_modules/react-native', :subspecs => [
    'Core',
    'DevSupport',
    'RCTText',
    'RCTNetwork',
    'RCTWebSocket',
    'RCTImage',
    'ART',
    'RCTActionSheet',
    'RCTGeolocation',
    'RCTPushNotification',
    'RCTSettings',
    'RCTVibration',
    'RCTLinkingIOS',
    'CxxBridge',
    'RCTAnimation',
    'RCTCameraRoll',
    'RCTBlob'
]
```

添加完成后执行以下命令进行安装。

```
pod install
```

此刻 iOS 工程中已经构建起了 React Native 环境,下一步就可以使用了。

```
// 初始化 Bridge,加载 JavaScript Bundle
RCTBridge *bridge=[[RCTBridge alloc] initWithDelegate:self launchOptions:
launchOptions];

// 开始创建视图:这里的 moduleName 需要与 index.js 的 registerComponent 中使用的名称一致
RCTRootView *rootView = [[RCTRootView alloc] initWithBridge:_bridge
```

```
                                          moduleName:@"example"
                                          initialProperties:nil];
[vc.view addSubview:rootView]
```

2. Android

首先在项目根目录的 build.gradle 文件中修改 maven 仓库地址配置,主要增加以下代码。

```
allprojects {
  repositories {
    ……
    maven {
      // 注意,如果React Native项目根目录不是Android目录的父级目录,则需要更改此处的相对路径
      url("$rootDir/……/node_modules/react-native/android")
    }
    maven {
      // 相对路径同上面保持一致
      url("$rootDir/……/node_modules/jsc-android/dist")
    }
  }
}
```

然后打开 module app 目录下的 build.gradle 文件,增加以下配置。

```
apply plugin: "com.android.application"

import com.android.build.OutputFile

project.ext.react = [
        entryFile   : "index.js", // 指定JavaScript文件入口
        enableHermes: false,  // clean and rebuild if changing
]

// 确保相对路径是准确的React Native node_modules路径
apply from: "../../node_modules/react-native/react.gradle"

def jscFlavor = 'org.webkit:android-jsc:+'

def enableHermes = project.ext.react.get("enableHermes", false);

android {
```

```
    splits {
        abi {
            reset()
            enable enableSeparateBuildPerCPUArchitecture
            universalApk false  // If true, also generate a universal APK
            include "armeabi-v7a", "x86", "arm64-v8a", "x86_64"
        }
    }

    applicationVariants.all { variant ->
        variant.outputs.each { output ->
            def versionCodes = ["armeabi-v7a": 1, "x86": 2, "arm64-v8a": 3, "x86_64": 4]
            def abi = output.getFilter(OutputFile.ABI)
            if (abi != null) {  // null for the universal-debug, universal-release variants
                output.versionCodeOverride =
                        versionCodes.get(abi)*1048576+defaultConfig.versionCode
            }

        }
    }

    packagingOptions {
        pickFirst '**/armeabi-v7a/libc++_shared.so'
        pickFirst '**/x86/libc++_shared.so'
        pickFirst '**/arm64-v8a/libc++_shared.so'
        pickFirst '**/x86_64/libc++_shared.so'
        pickFirst '**/x86/libjsc.so'
        pickFirst '**/armeabi-v7a/libjsc.so'
    }
}

dependencies {
    ......
    if (enableHermes) {
        // 确保相对路径是准确的React Native node_modules路径
        def hermesPath = "../../node_modules/hermesvm/android/";
        debugImplementation files(hermesPath + "hermes-debug.aar")
        releaseImplementation files(hermesPath + "hermes-release.aar")
```

```
    } else {
        implementation jscFlavor
    }
}

task copyDownloadableDepsToLibs(type: Copy) {
    from configurations.compile
    into 'libs'
}

// 确保相对路径是准确的React Native node_modules路径
apply from: file("../../node_modules/@react-native-community/cli-platform-
android/native_modules.gradle"); applyNativeModulesAppBuildGradle(project)
```

接下来打开Android项目根目录下的settings.gradle文件，增加以下配置：

```
// 确保相对路径是准确的React Native node_modules路径
apply from: file("../node_modules/@react-native-community/cli-platform-android/
native_modules.gradle"); applyNativeModulesSettingsGradle(settings)
```

最后单击Android Studio右侧的Gradle标签，选择项目主模块App（demo>app），单击Refresh Gradle Project按钮即可，如图1.12所示。

图1.12　Gradle管理面板

安装完依赖之后，需要在Android项目的Application.java中手动添加ReactNativeHost。

```
public class MainApplication extends Application implements ReactApplication {

    private final ReactNativeHost mReactNativeHost = new ReactNativeHost(this) {
        @Override
        public boolean getUseDeveloperSupport() {
            return BuildConfig.DEBUG;
        }
```

```java
    @Nullable
    @Override
    protected String getJSBundleFile() {
        // JavaScriptBundle存放路径，若返回null，则使用默认路径，getJSMainModuleName作为JavaScript文件名
        String jsBundleFile=getFilesDir().getAbsolutePath()+"/index.android.bundle";
        File file = new File(jsBundleFile);
        return file.exists() ? jsBundleFile : null;
    }

    @Override
    protected List<ReactPackage> getPackages() {
        // PackageList 类可以自动帮我们引入 package.json 安装的依赖
        List<ReactPackage> packageList=new PackageList(this).getPackages();
        return packageList;
    }

    @Override
    protected String getJSMainModuleName() {
        return "index"; // 指定 JavaScript 文件入口名称
    }
};

@Override
public ReactNativeHost getReactNativeHost() {
    return mReactNativeHost;
}

@Override
public void onCreate() {
    super.onCreate();
    // 初始化 React Native 依赖的 so 文件
    SoLoader.init(this, /* native exopackage */ false);
}
}
```

这样就搭建完成了Android的React Native环境。

1.4 本章小结

虽然React Native使得JavaScript开发者不用再关心iOS和Android的区别，能够用一套代码开发出接近于原生效果的App，但在实际业务场景中，如果熟悉原生开发的基本流程和相关概念，将大大降低协作成本，并有助于迅速排查问题。

第 2 章　React Native 启动流程及视图解析

上一章介绍了如何搭建一套 React Native 开发环境，并在原生 App（也称应用或应用程序）中看到了我们用 JavaScript 实现的 Hello World 视图。在这个过程中 React Native 究竟经历了哪些步骤，涉及哪些环节，而生成的原生视图又具备什么样的特性，这一章会给大家做一个详细的介绍。了解这些过程，可以帮助我们规避一些无法解决的异常写法，或者制定一些优化策略。

2.1　React Native 启动流程

在使用命令行构建 React Native（RN）项目后，可以获得两个完整的原生工程，分别位于 ios 和 android 目录。React Native 在这两个平台上的启动流程有所不同，但可以总结出大致相同的流程，如图 2.1 所示。

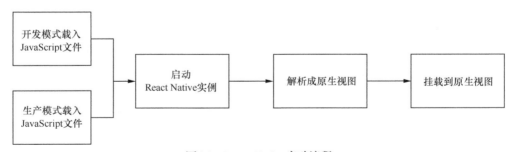

图 2.1　React Native 启动流程

如图 2.1 所示，我们之所以将载入的 JavaScript 文件拆分成开发模式和生产模式，是因为前者通过 Server（服务），也就是 IP+端口的方式载入 JavaScript 文件。此时文件尚未合并、压缩，开发者在工具中可以直接看到原始代码，方便后续调试，类似于 React 或 Vue 的开发模式；后者则载入了一个已经完成体积优化的 JavaScript 文件。

由于平台本身存在差异，因此下面我们会按 iOS 和 Android 分别描述具体的启动流程及涉及的概念。

2.1.1 iOS 启动流程

在默认项目 ios 目录下，启动流程代码基本都包含在 AppDelete.m 中：

```objc
- (BOOL)application:(UIApplication *)application didFinishLaunchingWithOptions:
(NSDictionary *)launchOptions
{
    RCTBridge *bridge = [[RCTBridge alloc] initWithDelegate:self launchOptions:
launchOptions];
    RCTRootView *rootView = [[RCTRootView alloc] initWithBridge:bridge
                                                   moduleName:@"example"
                                              initialProperties:nil];

    rootView.backgroundColor = [[UIColor alloc] initWithRed:1.0f green:1.0f
blue:1.0f alpha:1];

    self.window = [[UIWindow alloc] initWithFrame:[UIScreen mainScreen].bounds];
    UIViewController *rootViewController = [UIViewController new];
    rootViewController.view = rootView;
    self.window.rootViewController = rootViewController;
    [self.window makeKeyAndVisible];
    return YES;
}
```

即便你并不熟悉 iOS 原生代码，也可以根据类型和方法的名字识别出其中的基本元素和方法的效果。React Native 在 iOS 相关的类都是以 RCT 开头的，以下我们就介绍其中的几个关键类。

RCTBridge：可以理解为 React Native 在原生代码层的实例，iOS 可以使用该实例进行 React Native 的相关原生操作，例如页面渲染、事件传递、设置载入的 JavaScript 名称、实例重启等。

RCTRootView：利用 JavaScript 创建的原生视图，参数 moduleName 为我们在 React Native 端使用 AppRegistry.registerComponent 注册的组件名称，initialProperties 为向 JavaScript 传递的初始参数，在 React 组件的 props 中可以接收到。

UIApplication：一个 iOS App 创建的第一个对象就是 UIApplication 对象，并且这个对象是单例的。

UIViewController：iOS 最基础的视图控制器，我们可以简单地认为 iOS App 中的每一页都是一个 UIViewController。

iOS React Native 启动的整体流程大致可以解释成：UIApplication 创建后，RCTBridge 载入 React Native JavaScript 文件，根据 AppRegistry.registerComponent 注入的 moduleName 和 React 组件生成 RCTRootView，并且挂载到原生视图控制器（UIViewController）。

2.1.2　Android启动流程

相对而言，Android启动流程涉及的概念和类会多一些，默认项目android目录下，有两个文件MainApplication.java和MainActivity.java。

```java
public class MainApplication extends Application implements ReactApplication {

  private final ReactNativeHost mReactNativeHost = new ReactNativeHost(this) {
    @Override
    public boolean getUseDeveloperSupport() {
      return BuildConfig.DEBUG;
    }

    @Override
    protected List<ReactPackage> getPackages() {
      List<ReactPackage> packageList = new PackageList(this).getPackages();
      return packageList;
    }

    @Override
    protected String getJSMainModuleName() {
      return "index";
    }
  };

  @Override
  public ReactNativeHost getReactNativeHost() {
    return mReactNativeHost;
  }

  @Override
  public void onCreate() {
    super.onCreate();
    SoLoader.init(this, /* native exopackage */ false);
  }
}
```

在分析这段流程之前，有两个Android开发的原生概念——Application和Activity，需要和没有Android背景的开发者解释一下。

Application：每个Android App都有一个Application实例，在App开启的时候首先就会将它实例化，并且只实例化一次。

Activity：是 Android 中一个 App 组件，提供一个屏幕，用户通过与其交互来完成某项任务。我们在 App 上看到的每一"页"，就是一个 Activity。

MainApplication.java 中的 ReactNativeHost 是 React Native 实例在 Android 环境下的容器，等价于 iOS 中的 RCTBridge，也用于设置载入 JavaScript 文件路径、渲染页面等功能。当前示例中的 Application 实现了 ReactApplication 接口，确保在其他上下文中可以通过 getReactNativeHost 获取当前 React Native 的实例。

而在 MainActivity.java 中：

```java
public class MainActivity extends ReactActivity {
    /**
     * Returns the name of the main component registered from JavaScript.
     * This is used to schedule rendering of the component.
     */
    @Override
    protected String getMainComponentName() {
        return "example";
    }
}
```

这里主要重写了 getMainComponentName，确定了需要渲染的视图名称。我们如果进一步探究 ReactActivity.java 源码，会发现 getMainComponentName 最终被用于 ReactDelegate.java：

```java
mReactRootView.startReactApplication(
    getReactNativeHost().getReactInstanceManager(),
    appKey,    // 这个值即为 getMainComponentName
    mLaunchOptions);
```

该语句最终生成了一个由 React Native 绘制而成的 Android 原生视图，然后被挂载到了 Activity 中。Android 启动流程涉及的内部类较多，如图 2.2 所示。

图 2.2　React Native Android 视图相关类关系

ReactActivity：继承自 Android 原生视图（AppCompatActivity），主要包含了 ReactActivityDelegate 这个类的实例。

ReactActivityDelegate：包含了ReactDelegate的实例，大部分操作是通过调用ReactDelegate来完成的。

ReactDelegate：包含当前Activity实例，并创建了ReactRootView实例，通过ReactRootView来实现视图的创建和销毁。

ReactRootView：继承自Android原生视图类FrameLayout，也就是最后通过React Native JavaScript创建的原生视图实例。

为什么Android中使用了这么多代理来处理视图变化？不同于iOS，Android原生基础视图元素相对而言会多一些，例如Activity、Fragment和FragmentActivity等，这些基础视图包含的事件触发机制大同小异，如果直接操作ReactRootView，代码会相对冗余，类似的逻辑会散落在各个基础视图类中，所以利用代理的方式来统一处理ReactRootView，基础视图类只需调用代理的方法即可。

2.1.3 小结

iOS和Android中的基本概念和启动流程大体是一致的，可以分为以下3个步骤。

（1）创建当前平台下的React Native实例。

（2）使用React Native实例和JavaScript端注册的React组件来创建原生视图。

（3）将创建好的React Native原生视图挂载到当前平台屏幕上。

此时对于原生App来说，就是创建了一个原生视图，只不过创建方式稍有不同而已。

2.2 局部渲染React Native

上一小节描述了App整屏利用React Native渲染的情况，但在实际运用中，我们有时只需要在一屏的某个区域使用React Native，例如将一些需要快速响应变化的营销模块通过热更新机制快速上线。那么，如何将刚刚的整屏React Native视图变成局部视图呢？

2.2.1 iOS局部渲染

iOS的局部渲染相对简单，我们只需要修改RCTRootView实例的位置和大小即可。RCTRootView继承于iOS原生视图UIView，是最基础的视图控件，我们可以直接设置其属性frame来控制其位置和大小，具体示例如下。

```
// 创建Bridge和实例化RCTRootView
  RCTBridge *bridge = [[RCTBridge alloc] initWithDelegate:self launchOptions:nil];
  RCTRootView *rnRootView = [[RCTRootView alloc] initWithBridge:bridge moduleName:
@"example" initialProperties:nil];
```

```
// 设置RCTRootView的Frame，左边距0，上边距100，宽度同屏幕宽度，高度为140
CGFloat screenWidth = [UIScreen mainScreen].bounds.size.width;
rnRootView.frame = CGRectMake(0, 100, screenWidth, 140);

// 添加RCTRootView
[self.view addSubview:rnRootView];
```

最终效果如图2.3所示。

2.2.2　Android局部渲染

Android的局部渲染会涉及另一个原生概念——Fragment，可以把它理解为碎片，它主要有以下5个特点。

（1）Fragment是依赖于Activity的，不能独立存在。

（2）一个Activity里可以有多个Fragment。

（3）一个Fragment可以被多个Activity重用。

（4）Fragment有自己的生命周期，并能接收输入事件。

（5）在Activity运行时可动态地添加或删除Fragment。

React Native目前并不直接提供可实现局部渲染的类，但在master中却存在一个继承于Fragment的ReactFragment类，其中也包含了上节提到的ReactDelegate实例，来实际管理React Native实现的Fragment视图。所以如果需要实现局部渲染，开发者需要自己实现具体的视图逻辑。

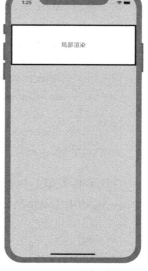

图2.3　iOS局部渲染

在Android环境中的局部渲染就是在原生的Activity中嵌入一个ReactFragment实例，并设置其位置和大小。那么要如何设置Fragment在Activity中的位置呢？这里就需要简单了解下Android中提供布局的xml方案。

我们先可以看这样一个Activity的布局xml示例：

```
<?xml version="1.0" encoding="utf-8"?>
<android.support.constraint.ConstraintLayout xmlns:android="http://schemas.
android.com/apk/res/android"
    xmlns:app="http://schemas.android.com/apk/res-auto"
    xmlns:tools="http://schemas.android.com/tools"
    android:layout_width="match_parent"
    android:layout_height="match_parent"
    tools:context=".C_2_2_2.Activity">
    <TextView
        android:id="@+id/textView"
```

```xml
        android:layout_width="wrap_content"
        android:layout_height="wrap_content"
        android:text="这是个原生的Activity"
        app:layout_constraintBottom_toBottomOf="parent"
        app:layout_constraintEnd_toEndOf="parent"
        app:layout_constraintStart_toStartOf="parent"
        app:layout_constraintTop_toTopOf="parent" />

    <FrameLayout
        android:id="@+id/fragment_container"
        android:layout_width="0dp"
        android:layout_height="300dp"
        app:layout_constraintBottom_toTopOf="@+id/textView"
        app:layout_constraintEnd_toEndOf="parent"
        app:layout_constraintStart_toStartOf="parent"
        app:layout_constraintTop_toTopOf="parent">

    </FrameLayout>
</android.support.constraint.ConstraintLayout>
```

和HTML类似，Android也是利用标签来确定各元素的关系。除了声明XML布局外，我们还需要在Activity中引入对应的XML，使屏幕和布局关联起来。

```java
@Override
protected void onCreate(Bundle savedInstanceState) {
    super.onCreate(savedInstanceState);
    setContentView(R.layout.activity_chapter2_2_2); // 使用xml中的布局绘制页面

    FragmentManager fragmentManager = getSupportFragmentManager();
    if (fragmentManager.findFragmentById(R.id.fragment_container) == null) {
        ReactFragment.Builder fragmentBuilder = new ReactFragment.Builder();
        fragment = fragmentBuilder
            .setComponentName("2_2_1") // 设置JavaScript端注册的ComponentName
            .setLaunchOptions(new Bundle()) // 设置传入的初始参数
            .build();

    } else {
        fragment = (ReactFragment) fragmentManager.findFragmentById(R.id.fragment_container);
    }
```

```
fragmentManager
    .beginTransaction()
    .add(R.id.fragment_container, fragment)
    .commit();
}
```

最终效果如图2.4所示。

图2.4　Android局部渲染

2.3　React Native原生视图详解

在上面几节的代码中我们已经看到了React Native为各平台提供的原生UI视图类,那么其中具体会包含哪些属性和方法,我们又可以怎样合理利用它们呢?本节会和大家详细地解读这两个问题。

2.3.1　iOS——RCTRootView

RCTRootView继承自iOS的基础视图UIView,它本质是窗口上的一块区域,也是iOS中所有控件的基类,负责内部区域的渲染、触摸事件、动画,也可以管理本身所有的子视图。

RCTRootView提供了两种初始化的方法，initWithBridge和initWithBundleURL，前者利用已经实例化的RCTBridge生成视图，后者则直接使用JavaScript的本地地址生成RCTBridge后再生成视图。通常情况下，我们会采用initWithBridge以避免重复生成RCTBridge，减少性能消耗。而在初始化的过程中，为了避免白屏，RCTRootView允许我们重写其中的loadingView属性，在React Native视图尚未显示出来时，显示一个等待中的视图，并可设置这个等待视图的移除时间和动画。

比如，我们可以在2.1.1小节的例子中增加一个loading状态。

```
// 创建Bridge和实例化RCTRootView
  RCTBridge *bridge = [[RCTBridge alloc] initWithDelegate:self launchOptions:nil];
  RCTRootView *rnRootView = [[RCTRootView alloc] initWithBridge:bridge moduleName:
@"example" initialProperties:nil];

  // 设置loadingView的控件，这里使用iOS默认的loading来作为loadingView
  UIActivityIndicatorView *loadingView = [[UIActivityIndicatorView alloc] init
WithActivityIndicatorStyle:UIActivityIndicatorViewStyleGray];
  [loadingView startAnimating];
  rnRootView.loadingView = loadingView;

  // 设置loadingView消失的动画持续时间（默认为0.25s）
  rnRootView.loadingViewFadeDuration = 0.3;

  // 设置内容加载完成后的loadingView延时展示的时长，当loadingViewFadeDuration为0时，此属性不生效
  rnRootView.loadingViewFadeDelay = 10;
  self.view = rnRootView;
```

另外，在事件处理方面，RCTRootView则提供了cancelTouches，能够在视图内部取消JavaScript的手势事件，避免在某些场景下原生手势和JavaScript手势冲突。

2.3.2 Android——ReactRootView

ReactRootView继承自Android的FrameLayout（帧布局），是最简单的界面布局，默认把元素放在屏幕的左上角，后续添加的元素会覆盖前一个，如果元素一样大，那么同一时刻只能看到最上面的那个元素。

我们可以通过图2.5简单了解Android的布局种类和具体实现的基类。

具体实现布局的基类主要是ViewGroup和View。

ViewGroup：放置View的容器，给childView计算出建议宽、高和测量模式，以及决定childView的位置。

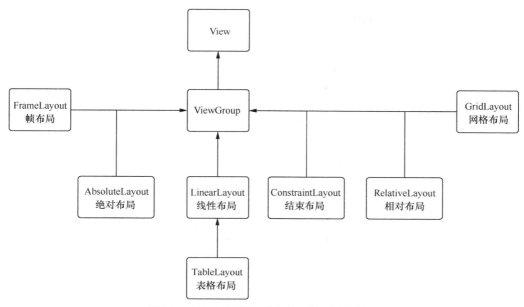

图2.5 Android局部方式及各布局之间的关系

View：根据测量模式和ViewGroup给出的建议宽和高，计算出自己的宽和高，并且在ViewGroup为其指定的区域内绘制自己的形态。

了解了ReactRootView的布局机制后再来分析具体的源码，其内部还包含以下主要属性及方法。

mReactInstanceManager：ReactInstanceManager实例，在startReactApplication调用的时候传入。在大部分应用场景下，我们会使用ReactNativeHost来获取ReactInstanceManager实例，而ReactInstanceManager则是React Native用于管理ReactContext的模块。需要额外说明的是，在Android中，Context（上下文）描述了当前使用者的场景及资源。例如，React Native的ReactContext就包含了CatalystInstance（处理JavaScript和Java接口的相互调用）、UI、原生模块、JavaScript通信的消息队列线程，以及当前活动的Activity等描述当前场景的属性。

mCustomGlobalLayoutListener：实现了ViewTreeObserver.OnGlobalLayoutListener接口，用于监听视图树，当视图树发生变化时，会通知到当前视图。

mJSTouchDispatcher：JSTouchDispatcher实例，负责将视图接收到的Touch事件传递给JavaScript端。

startReactApplication：创建ReactContext，建立视图监听，并绘制React Native组件。

ReactRootView整体的渲染流程，以及相关的类和方法如图2.6所示。

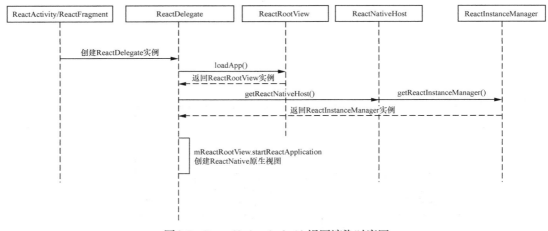

图 2.6　React Native Android 视图渲染时序图

2.3.3　视图长度单位

在实际开发中，我们会使用 React Native 提供的 StyleSheet 给视图绘制样式，通过 StyleSheet.create() 创建视图的宽/高、边距等尺寸。

```
const styles = StyleSheet.create({
  container: {
    width: 100,
    height: 100,
    paddingHorizontal: 10,
    ......
  },
});
```

那么 React Native 究竟怎么处理这些无单位的尺寸的大小？在此之前，我们先总结一下移动端的各种尺寸单位及其含义。

px：像素，是图像的基本采样单位。

DPI/PPI：都表示像素密度，也就是每英寸（in,1in=25.4mm）的像素点数。

pt：iOS 常用单位，独立像素，代表一个绝对长度，不随屏幕像素密度的变化而变化。

dp：Android 常用单位，设备无关像素（device independent pixels），这种尺寸单位在不同设备上的物理大小相同。

可以看出，iOS 的 pt 单位和 Android 的 dp 单位的定义基本类似，都是为了避免像素密度的影响，让开发者无须关注屏幕密度、物理像素之间的换算关系。所以，React Native 也将自己的尺寸单位根据平台转化成了这两种单位。

另外，React Native 也提供了 PixelRatio 对象，开发者可以直接访问设备的像素密度。该对象提供以下几种方法。

get：返回设备的像素密度。常见的设备像素密度如下。

1：Android 的 mdpi 设备（160dpi）。

1.5：Android 的 hdpi 设备（240dpi）。

2：最常见的设备，包括 iPhone 4～8（除 p 系列），Android 的 xhdpi 设备（320dpi）。

3：iPhone6～8 的 p 系列，iPhone X/XS/XS Max，Android 的 xxhdpi 设备（480dpi）。

3.5：Nexus 6 等少数设备。

getFontScale：返回字体缩放比例，如果没有设置字体缩放，它会直接返回设备的像素密度。目前这个函数仅仅在 Android 设备上实现了，在 iOS 设备上它会直接返回默认的像素密度。

getPixelSizeForLayoutSize：将一个布局尺寸（dp）转换为像素尺寸（px），且返回的一定是整数。

roundToNearestPixel：返回最接近对齐物理像素的设备独立像素值。通常用于 px 转化成 dp，例如 UI 设计图是以 px 为单位，但需要在不同屏幕上等比缩放，那我们就可以设计一个通用转换算法。

```
import {Dimensions} from 'react-native';
const deviceWidthDp = Dimensions.get('window').width;
// UI 设计图的宽度为 640
const widthPx = 640;
function pxToDp(elePx) {
  return elePx * deviceWidthDp / widthPx;
}
```

除了无单位的数值外，React Native 也逐渐支持使用百分比（0.42 版本之后），可以用于描述 width、height、top、left 等属性。

```
export default class App extends Component {
  render() {
    return (
      <View style={{ height: '100%' }}>
        <View style={{ width: '25%', height: '25%', top: '50%', left: '50%',
borderWidth: 1 }}>
          <Text>25%</Text>
        </View>
      </View>
    );
  }
}
```

实际效果如图2.7所示。

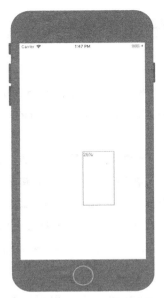

图2.7　React Native绝对定位

2.4　React Native布局方式

React Native中常见的布局方式是Flex布局和绝对定位布局。Flex布局是2009年由W3C提出的布局方案，目前在浏览器端已经得到了广泛支持。对Web开发者来说，Flex布局和绝对定位布局很简单；而对原生开发者而言，这两种布局方式可能会有点陌生，我们会在本节对这两种布局方式进行一个具体的说明。

2.4.1　Flex布局

Flex布局是Flexible Box的缩写，意为"弹性布局"，主要分为Flex Container（容器）和Flex item（子元素）。容器默认存在两根轴，即默认水平的主轴和默认垂直的次轴。

下面就是一个Flex布局的常见场景，布局效果如图2.8所示。

```
<View style={{ display: 'flex', flexDirection: 'row', paddingHorizontal: 10 }}>
  <View style={{ flex: 1, borderWidth: 1, height: 100 }}>
    <Text>item 1</Text>
  </View>
  <View style={{ flex: 1, borderWidth: 1, height: 100 }}>
    <Text>item 2</Text>
  </View>
```

```
    <View style={{ flex: 1, borderWidth: 1, height: 100 }}>
      <Text>item 3</Text>
    </View>
</View>
```

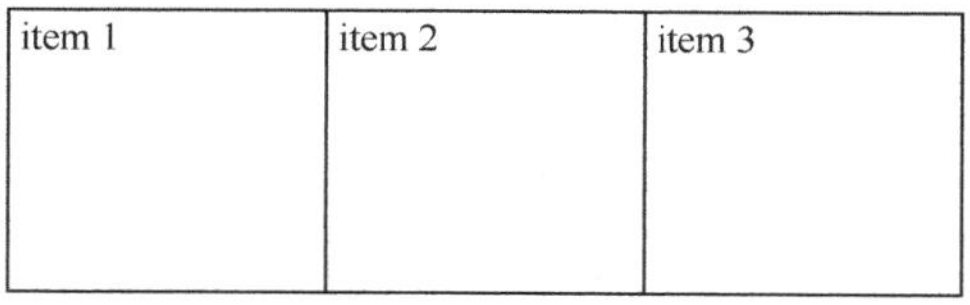

图2.8 Flex布局效果

Flex容器主要有以下几个属性。

flexDirection:决定主轴方向,值可以为以下4种。

- row:默认值,主轴为水平方向,起点在左端。
- row-reverse:主轴为水平方向,起点在右端。
- column:主轴为垂直方向,起点在上沿。
- column-reverse:主轴为垂直方向,起点在下沿。

flexWrap:默认情况下,flex item都排在一条轴线上,flex-Wrap可决定如果一条轴线排不下,如何换行。

- nowrap:不换行。
- wrap:换行,从主轴起点开始。
- wrap-reverse:换行,从主轴末尾开始。
- justifyContent:定义了flex item在主轴上的对齐方式,具体效果如图2.9所示。
- flex-start:左对齐。
- flex-end:右对齐。
- center:居中。
- space-between:两端对齐,item之间间隔相等。
- space-around:item两侧的间隔相等。
- alignItems:定义了flex item在次轴上的对齐方式,具体效果如图2.10所示。
- flex-start:次轴的起点对齐。
- flex-end:次轴的终点对齐。
- center:次轴的中点对齐。
- baseline:item的第一行文字基线对齐,常用在不同字号对齐的场景下。
- stretch:默认值,如果item在次轴方向没有设置固定尺寸,将占满整个容器。
- alignContent:定义了多根轴线的对齐方式,如果主轴只存在一根轴线,该属性不起作用。
- flex-start:与次轴的起点对齐。

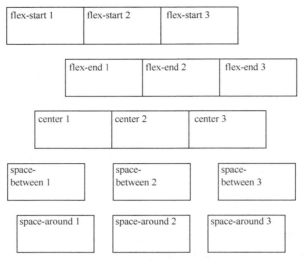

图2.9 justifyContent不同值的效果

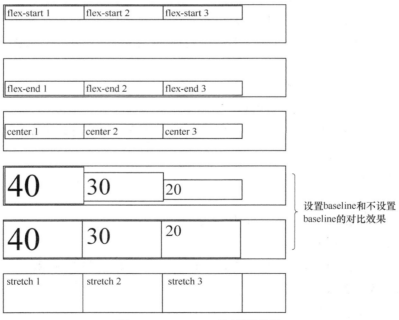

图2.10 alignItems不同值的效果

- flex-end：与次轴的终点对齐。
- center：与次轴的中点对齐。
- space-between：与次轴两端对齐，轴线之间的间隔平均分布。
- space-around：每根轴线两侧的间隔都相等。

- stretch：轴线占满整个次轴。

作为子元素的item，包含以下属性。
- flexGrow：item的放大比例，默认为0，即如果存在剩余空间，也不放大。
- flexShrink：item的缩小比例，默认为1，即如果空间不足，该item将缩小。
- flexBasis：在分配多余空间之前，item占据的主轴空间，计算主轴是否有多余空间。它的默认值为auto，即项目的本来大小。
- flex：flex属性是flexGrow、flexShrink和flexBasis的简写，默认值分别为0、1、auto，后两个属性为可选。其最常见的用法是按flexGrow比例分配控件，或部分item确定长度，剩余的item填充空间。flex与浏览器中CSS的区别在于在React Native中flex只能为整数值。
- flex为正整数，组件尺寸有弹性，并根据具体的flex值按比例分配。
- flex为0，组件尺寸由width和height决定且不再有弹性。
- flex为−1，组件尺寸一般还是由width和height决定。但是当空间不够时，组件尺寸会缩小到minWidth和minHeight所设定的值。
- alignSelf：允许单个item有不一样的对齐方式，可覆盖align-items属性。默认值为auto，表示继承父元素的align-items属性，如果没有父元素，则等同于stretch。

在React Native中，Yoga是用于实现跨平台布局系统的引擎，遵守W3C规范，兼容Flex布局，支持Java、C#、Objective-C和C 4种语言。其底层代码使用C语言编写，性能良好，并且很容易跟其他平台集成。除了React Native之外，Facebook在自己的原生UI渲染框架中也使用了这个引擎，例如Android的Litho，iOS的ComponentKit。

如图2.11所示Yoga提供了一个在线的playground，可用于直接调试Flex布局，修改元素个数、布局容器属性和item属性，并且可以直接生成对应平台（包括React Native、Android和iOS）的代码。

图2.11 Yoga playground

2.4.2 绝对定位

除了Flex布局外，React Native中也包含了Web开发者常用的绝对定位布局，也就是通过设定元素的top、left、right和bottom来确定元素的位置。与Web不同的是，React Native中每个元素的position属性都默认为relative，也就是说每个position设置为absolute的元素，它的top、left、right和bottom都是相对于自己的父元素的位置，例如：

```
<View style={{ borderWidth: 1, height: 500 }}>
    <View style={{ position: 'absolute', top: 100, left: 100, width: 100, height: 100, borderWidth: 1 }}>
        <Text>item 1</Text>
    </View>
    <View style={{ position: 'absolute', top: '50%', left: '50%', width: 100, height: 100, borderWidth: 1 }}>
        <Text>item 2</Text>
    </View>
</View>
```

在使用绝对定位布局时（见图2.12），通常有以下两个方面需要特别考虑。

（1）层级关系。在CSS中，通常使用z-index来控制图层的层级关系，z-index数值越大，图层越高。在React Native中，除了zIndex外，还有一个Android属性elevation，也会影响到视图的层级关系。elevation使用了视图高度来决定应用的视图效果。因此同时使用zIndex和elevation的时候，就需要注意其中的层级影响。

没有zIndex，没有elevation：由自身结构决定，结构下面的视图在上层。

有zIndex，没有elevation：zIndex数值大的在上层。

没有zIndex，有elevation：elevation数值大的在上层。

有zIndex，有elevation：以elevation为准。

我们可以参考下面这个例子：

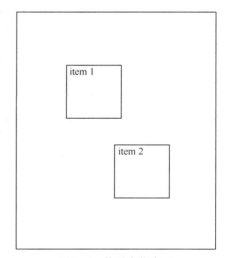

图2.12 绝对定位布局

```
<View style={{ borderWidth: 1, height: 500 }}>
    <View style={{ position: 'absolute', top: 0, left: 0, width: 100, height: 100, borderWidth: 1, zIndex: 10, backgroundColor: 'white' }}>
```

```
        <Text>only zIndex 10</Text>
    </View>
    <View style={{ position: 'absolute', top: 25, left: 25, width: 100, height:
100, borderWidth: 1, zIndex: 15, backgroundColor: 'white' }}>
        <Text>only zIndex 15</Text>
    </View>
    <View style={{ position: 'absolute', top: 50, left: 50, width: 100, height:
100, borderWidth: 1, backgroundColor: 'white', elevation: 15 }}>
        <Text>only elevation 15 </Text>
    </View>
</View>
```

具体效果如图2.13所示,左侧为在iOS系统中的显示效果,右侧为在Android系统中的显示效果。

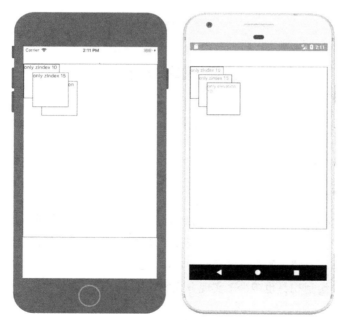

图2.13 iOS和Android中zIndex和elevation的显示效果

(2) overflow (超出显示、溢出)。在开发中,有时需要制作子视图超出父视图的场景,例如在消息提醒之类的图标右上角展示小红点。在CSS中,通过设置overflow:hidden或overflow:visible来控制父视图是否允许显示子视图超出的部分。在Android中,React Native的overflow:visible并不能生效,也就是说子视图不能超出父视图显示。当然,也不是完全没有办法,Android提供了相关属性来支持这样的操作:给对应视图A设置android:clipChildren="false",这样该布局下的子视图B允许自己的子视图C超出自己显示(视图C是视图A的孙子级视图),然后通过React Native提供的创建自定义原生UI组件的方式（第8章会详细讲解）将这个能力提供给JavaScript端，使子视图具备超出父视图显示的能力，实际效果如图2.14所示。

图2.14　Android中实现超出/显示效果

2.5　本章小结

本章主要解释了React Native在各平台上的启动流程，以及最终转化成的原生组件。了解这些过程及原生视图的特点，有助于我们更好地优化流程，采取预加载、缓存等策略缩短首屏渲染时间，以提供更好的用户体验。相对统一的布局方式在一定程度上减少了跨端绘制页面的成本，三端的开发者都采取相同的布局思维绘制UI，也减少了实现方式的差异。

第3章 文本及输入

文本承载着页面上大部分的内容展示，可以说文本是除了基础的视图单元之外，运用得最广的元素之一。本章会给大家详细描述 React Native 是如何处理文本展示及文本输入的，并存在哪些需要注意的使用方式。

3.1 Text 解析

在 React Native 中，可以使用 <Text> 组件展示文本信息，方式非常简单、直接，例如：

```
<Text>Hello React Native</Text>
```

组件本身也支持样式及触摸事件，例如：

```
<Text
  style={{ fontSize: 20, color: '#333333' }}
  onPress={() => Alert.alert('touched')}
>
  Hello React Native
</Text>
```

从布局上讲，Text 组件没有类似于 CSS 行内元素这样的概念，所以单个 Text 组件即为一行，但它属于 Flex 布局范畴，可以使用 flexDirection 属性设置行内并列的效果，例如：

```
<View style={{ flex: 1, justifyContent: 'center' }}>
  <View>
    <Text style={{ fontSize: 40, borderWidth: 1 }}>1</Text>
    <Text style={{ fontSize: 40, borderWidth: 1 }}>2</Text>
    <Text style={{ fontSize: 40, borderWidth: 1 }}>3</Text>
  </View>
  <View style={{ flexDirection: 'row' }}>
```

```
        <Text style={{ fontSize: 40, borderWidth: 1 }}>1</Text>
        <Text style={{ fontSize: 40, borderWidth: 1 }}>2</Text>
        <Text style={{ fontSize: 40, borderWidth: 1 }}>3</Text>
    </View>
</View>
```

效果如图3.1所示。

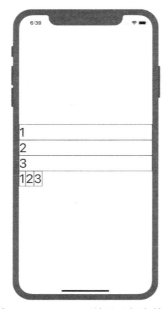

图3.1 使用flexDirection属性设置行内并列的效果

除了这些最基础的展示外,Text还提供了另外一些常用的属性。

selectable:决定用户是否可以长按选择文本,以便复制和粘贴。

ellipsizeMode:当文本内容超出当前Text组件大小时,使用省略号对其进行截取的方式,主要包含以下4种模式。

(1) head:从文本内容头部进行截取。

(2) middle:从文本内容中间进行截取。

(3) tail:从文本内容尾部进行截取。

(4) clip:不显示省略号,直接不显示超出内容。

numberOfLines:设置文本行数,若内容超过行数限制,根据ellipsizeMode的设定进行裁剪。

allowFontScaling:控制字体是否要根据系统的"字体大小"进行缩放。

需要注意的是,ellipsizeMode的head模式仅作用在Text最后一行的头部,并且head和middle模式在Android中不生效,例如:

```
  <Text style = {{ width: 200, fontSize: 40, borderWidth: 1 }} ellipsizeMode =
{'head'} numberOfLines = {2} >
      文字过长文字过长文字过长文字过长
  </Text>
  <Text style = {{ width: 200, fontSize: 40, borderWidth: 1 }} ellipsizeMode =
{'middle'} numberOfLines = {2} >
      文字过长文字过长文字过长文字过长
  </Text>
  <Text style = {{ width: 200, fontSize: 40, borderWidth: 1 }} ellipsizeMode =
{'tail'} numberOfLines = {2} >
      文字过长文字过长文字过长文字过长
  </Text>
```

实际效果如图3.2所示。

图3.2　iOS和Android下ellipsizeMode效果

众所周知，React Native会将由JavaScript编写的React组件转化成对应的原生UI组件，所以Text组件实际的功能也都受相关平台的影响，下面将讲解Text组件具体的实现过程及原理。

3.1.1　RCTTextView和ReactTextView

RCTTextView和ReactTextView分别是Text组件在iOS和Android上对应的原生UI组件，两者在结构和实现上都存在一定的差异，所以下面将分平台描述其具体的实现。

1. iOS

UILabel可以说是iOS开发中除UIView之外使用得最广泛的控件之一，其自身也继承于UIView，最简单的用法就是展示文字。开发者只需新建一个UILabel对象并设置其text值，就能得到一段展示文本。当然，UILabel自身也有很多属性可以设置，例如字体大小、颜色、高亮等。但React Native并没有把自身的Text组件解析成UILabel，而是使用了自己实现的类RCTTextView，大致结构如图3.3所示。

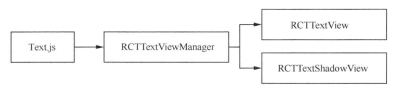

图3.3　Text组件与iOS原生类关系

RCTTextViewManager：几乎所有的React Native视图组件都会包含一个Manager管理类，用来声明可被JavaScript使用的属性和方法（具体用法会在第8章自定义原生组件中详细说明），RCTTextViewManager就是RCTTextView的管理类。RCTTextViewManager除了包含RCTTextView之外，还包含RCTTextShadowView。

RCTTextView：React Native Text组件实际转化成的类，包含NSTextStorage（文字存储器）、RCTTextRenderer（文字渲染引擎）和CALayer/CAShapeLayer（展示图层）等相关属性。

RCTTextShadowView：类似于React中Virtual Dom的概念，每次更新时会先检测该实例是否需要更新，如果需要才会更新对应RCTTextView。

另外，RCTTextView中的文本渲染类RCTTextRenderer使用了性能更好的CALayer和CATiledLayer图层类来进行具体的实现，大致可以这么简单理解。

CALayer/CATiledLayer：作为iOS负责内容展示的基类，但并不参与用户点击事件的处理。

RCTTextRenderer：用作CALayer的delegate，重写了CALayer中用于显示内容的drawLayer:inContext方法，也就是直接定义了文本的渲染规则。

由于iOS是闭源的系统，我们无法探知UILabel是如何实现文字展示的，因此React Native采用了这种自定义的方式来实现具体的文本渲染逻辑，并对一些特殊场景进行了性能优化。例如对超出屏幕1.5倍的大段文本采用了CATiledLayer代替CALayer展示。CATiledLayer的绘制任务是在非主线程中进行的，不会因为阻塞主线程而造成视觉上的卡顿，且只渲染屏幕范围内的可见内容，不会造成内存跟随需要渲染的内容线性增长，通常还会用在展示大屏图片上。

2. Android

在Android系统中，我们通常会使用TextView设置文字展示，可以在XML布局文件中指定属性，也可以在Java文件中动态设置（对于JavaScript开发者而言，这一过程类似于手动修改DOM元素属性值）。ReactTextView继承于TextView，具体如图3.4所示。

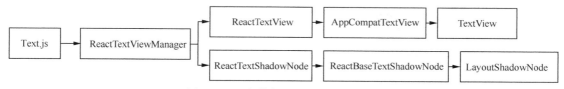

图3.4　Text组件与Android原生类关系

ReactTextViewManager：ReactTextView管理类，主要包括ReactTextView和ReactTextShadowNode，以及它们的更新机制。

ReactTextView：具体实现文本的类，重写了onLayout方法，用于计算自身的位置，主要负责实际的渲染。

ReactTextShadowNode：继承于ReactBaseTextShadowNode，通过Yoga实现内容的拼接及具体的更新机制。

从整体的流程上说，JavaScript端定义的Text结构会被转化为ReactTextShadowNode的树形结构，然后进行内容拼接和样式继承，之后将更新类ReactTextUpdate加入UIViewOperationQueue中，等待队列执行，通知ReactTextView进行视图的渲染和更新。

3.1.2　行间距

行间距是处理文字排版，特别是多行文字排版中非常重要的概念。各平台对于行间距的定义和设置各不相同，以中文排版来说，设计师通常关心的是上一行文字的底部与下一行文字的顶部之间的距离（行间距），如图3.5所示。

图3.5　行间距

但在实际开发中，这一段距离通常不能被直接设置，文字内容也并不会占满文字组件的整个高度，所以通常是好几个属性影响这一段的高度。例如，在iOS中这一距离受字体行高font.lineHeight和行间距lineSpacing的影响，而在Android的TextView中则是受android:lineSpacingExtra（行间距）或android:lineSpacingMultiplier（行间距倍数）的影响。

在React Native中，则定义了类似于CSS的line-height样式属性来控制行间距，例如：

```
<View style={{ flexDirection: 'row' }}>
  <Text style={{ backgroundColor: 'black', color: 'white', fontSize: 28,
lineHeight: 40 }}>
```

```
    我是一段文字
  </Text>
  <View style={{ backgroundColor: '#666666', height: 42 }}>
    <Text style={{ color: 'white' }}>42</Text>
  </View>
  <View style={{ backgroundColor: '#999999', height: 40 }}>
    <Text style={{ color: 'white' }}>40</Text>
  </View>
</View>
```

实际效果如图3.6所示。

图3.6　文字实际高度（1）

从图3.6中可以看出，我们给一段文字设置了40的行高，但整体组件的高度却与高度为42的视图持平。这里的组件高度是由上文提到的Facebook的Yoga布局框架计算得到的，在React Native中，每个最小视图单元都是一个Yoga Node，它自身会包含一套计算位置与大小的算法。Text组件高度除了受行高的影响外，也会受输入内容的字符集影响，例如：

```
<View style={{ flexDirection: 'row', marginTop: 10 }}>
  <View>
    <Text style={{ backgroundColor: 'black', color: 'white', fontSize: 28 }}>我是一段文字</Text>
  </View>
  <View style={{ backgroundColor: '#999999', height: 30 }}><Text style={{ color: 'white' }}>30</Text></View>
</View>
  <View style={{ flexDirection: 'row', marginTop: 10 }}>
  <View>
    <Text style={{ backgroundColor: 'black', color: 'white', fontSize: 28 }}>我是一段文字agl</Text>
  </View>
  <View style={{ backgroundColor: '#999999', height: 34 }}><Text style={{ color: 'white' }}>34</Text></View>
</View>
```

实际效果如图3.7所示。

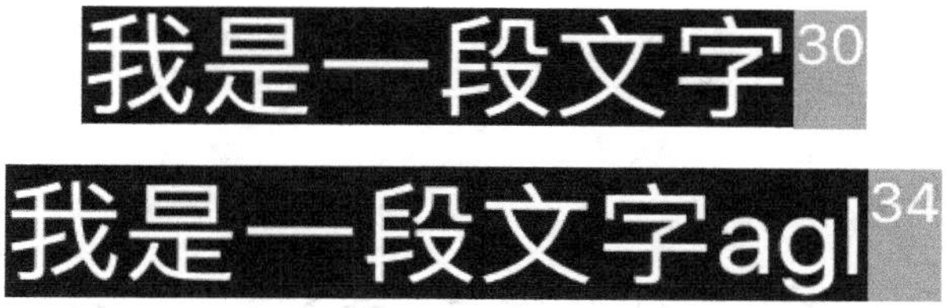

图3.7 文字实际高度（2）

可以看出，同样是字号为28的文字，由于内容不一致，Text的高度会有变化，从而影响行间距的实际高度。

3.2 Text布局方式

除了已经介绍过的Flex布局、Text组件嵌套文字外，React Native还提供了Text组件的其他用法，使之能更符合Web开发者的使用习惯。当然，也会出现一些意料之外的效果，我们需要采取适当的用法，在满足需求的同时规避一些已知的问题。

3.2.1 Text的嵌套

Text的嵌套主要是为了满足富文本的处理，在一些信息展示类的场景中，通常需要将同一段落中的部分文字的字号、颜色另外设值，以达到视觉上的区分。在React Native中可以这样处理：

```
<Text style={{ fontSize: 28, color: '#999999' }}>
    我是一段普通文字
    <Text style={{ color: '#333333' }}>我是一段醒目的文字</Text>
</Text>
```

实际效果如图3.8所示。

图3.8 文本嵌套

我们可以利用类似于HTML中Span标签的写法实现一段文字中包含不同样式的需求。嵌套的Text组件支持样式继承，内部的Text能够使用外部Text设置的style样式。另外，React Native除了可以嵌套Text组件，还可以嵌套Image（图片）组件，例如：

```
<Text style={{ fontSize: 28, borderColor: '#333333', borderWidth: 1 }}>
    <Image source={require('./icon.png')} style={{ height: 24, width: 24 }} />
    我是一段普通文字
</Text>
```

实际效果如图3.9所示。

✓ 我是一段普通文字

图3.9 图片文本嵌套

在iOS中，嵌套的Text组件会将内部的文字内容和样式转化为NSTextStorage，该类继承于NSMutableAttributedString，它本身就是iOS开发者常用的富文本类，可以携带多个文字样式，并且支持嵌套图片。

而Android的TextView则使用了支持显示富文本字符串Spannable对象，JavaScript端会遍历嵌套的Text组件，将内容拼接成一个SpannableString对象，并通过分段设置Span样式，来实现嵌套和样式继承。

不过React Native中的Text嵌套写法也存在以下3种局限性。

（1）被嵌套组件与位置相关的style样式几乎都不生效。

```
<Text style={{ fontSize: 28 }}>
    我是一段普通文字
    <Text style={{ paddingLeft: 10 }}>左Padding 10</Text>
    <Text style={{ marginLeft: 10 }}>左Margin 10</Text>
</Text>
```

实际效果如图3.10所示，可以看出，内嵌的Text组件样式paddingLeft和marginLeft均未生效。

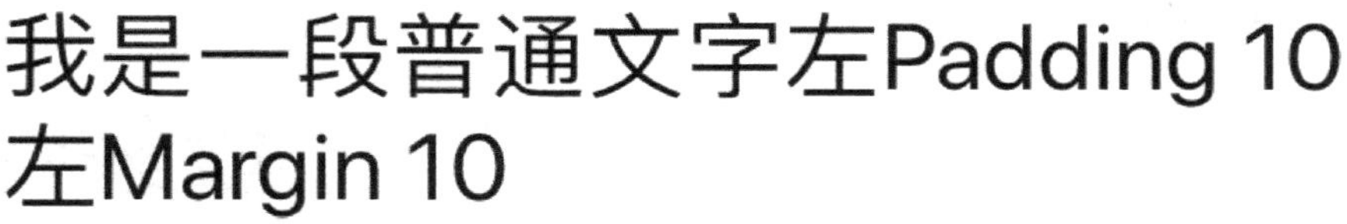

我是一段普通文字左Padding 10 左Margin 10

图3.10 文本效果（1）

（2）内嵌非Text组件会导致整体Text的lineHeight失效。

```
<Text style={{ fontSize: 28, borderColor: '#333333', borderWidth: 1, lineHeight: 40 }}>
    <Image source={require('./icon.png')} style={{ height: 24, width: 24 }} />
    行高失效
```

```
</Text>
<Text style={{ fontSize: 28, borderColor: '#333333', borderWidth: 1, lineHeight: 40 }}>
    行高<Text>有效</Text>
</Text>
```

实际效果如图3.11所示。

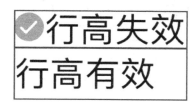

图3.11　文本效果（2）

（3）内嵌Text的numberOfLines属性会失效。

```
<Text style={{ fontSize: 28, borderWidth: 1 }}>
    1. <Text numberOfLines={2} ellipsizeMode={'tail'}>我是一段普通文字我是一段普通文字我是一段普通文字我是一段普通文字我是一段普通文字</Text>
</Text>
```

实际效果如图3.12所示。

图3.12　文本效果（3）

3.2.2　同行多字号文本的对齐方式

同行多字号文本的对齐也是很常见的需求，处理方式如下：

```
<View>
    <Text style={{ fontSize: 20, borderWidth: 1 }}>
        我是文字<Text style={{ fontSize: 30 }}>我是大一点的文字</Text>
    </Text>
</View>
<View style={{ flexDirection: 'row', marginTop: 10, borderWidth: 1 }}>
```

```
    <Text style={{ fontSize: 20 }}>我是文字</Text>
    <Text style={{ fontSize: 30 }}>我是大一点的文字</Text>
</View>
```

实际效果如图3.13所示。

图3.13　文本效果（4）

对于嵌套Text组件来说，其内部遵循的是原生的文本布局方式，不是Flex布局，相比Web开发者习惯的CSS语法，React Native在文本的布局和对齐方式上缺失了很多功能。另外，在单个Text中生效的部分属性会在嵌套过程中失去作用，例如仅在Android中生效的textAlignVertical（用于控制文字在垂直方向的布局的属性），在嵌套Text中不起作用。例如：

```
<View style={{ flexDirection: 'row', marginTop: 10, borderWidth: 1 }}>
    <Text style={{ fontSize: 20, textAlignVertical: 'center' }}>我是文字</Text>
    <Text style={{ fontSize: 30 }}>我是大一点的文字</Text>
</View>
<View style={{ marginTop: 10 }}>
    <Text style={{ fontSize: 20, borderWidth: 1, textAlignVertical: 'top' }}>
        我是文字<Text style={{ fontSize: 30 }}>我是大一点的文字</Text>
    </Text>
</View>
```

在iOS中的实际效果如图3.14所示。

图3.14　文本效果（5）

在Android中的实际效果如图3.15所示。

图3.15 文本效果（6）

如果使用不同的Text组件设置不同字号，那么对齐的方式仍然是使用Flex布局对齐，例如垂直居中：

```
<View style={{ flexDirection: 'row', marginTop: 10, borderWidth: 1, alignItems: 'center' }}>
    <Text style={{ fontSize: 20 }}>我是文字</Text>
    <Text style={{ fontSize: 30 }}>我是大一点的文字</Text>
</View>
```

不过需要注意的是，由于字号大小不一，小字号文字的上间距会略小，需要利用padding进行微调，例如将上例中的alignItems值改为flex-start，实际效果如图3.16所示。

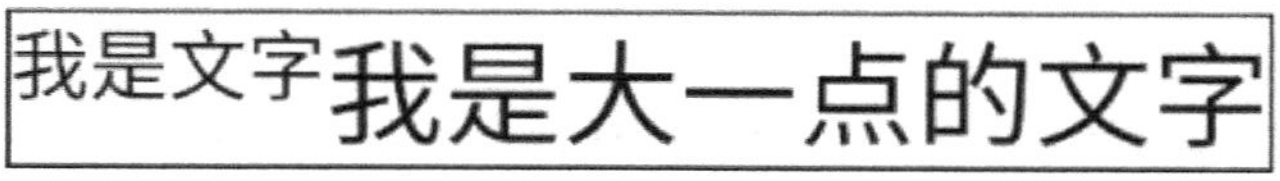

图3.16 文本效果（7）

3.3 文本输入——TextInput

React Native 内置了TextInput作为处理用户输入的基本组件，提供了自动获取焦点、指定键盘类型等常用功能，最基本的使用方式如下：

```
<TextInput
    style={{ width: 300, height: 40, borderWidth: 1 }}
    value={this.state.input}
    onChangeText={text => this.setState({ input: text })}
/>
```

实际效果如图3.17所示。

需要注意的是，TextInput在Android中默认有一个底边框且存在内边距。如果想让它看起来和iOS上的效果尽量一致，则需要将padding的值设置为0。

与Text组件类似，TextInput也存在它本身的JavaScript组件、各平台的管理类，以及具体实现使用的原生类或自定义原生类。

图3.17　输入框效果

1. iOS

iOS中处理文本有两个原生类，即UITextField和UITextView，分别用于处理单行和多行文本输入，React Native分别重写了对应的RCTUITextField和RCTUITextView，内部通过RCTBackedTextInputDelegateAdapter代理调用了大部分原生UI的方法。TextInput组件具体的iOS原生类如图3.18所示。

图3.18　TextInput组件iOS原生类

RCTMultilineTextInputViewManager和RCTSinglelineTextInputViewManager分别是RCTMultilineTextInputView和RCTSinglelineTextInputView对应的组件管理类，用于实现组件属性导出和组件的实例化。

RCTMultilineTextInputView和RCTSinglelineTextInputView则主要管理RCTUITextView和RCTUITextField组件，包括默认的属性设置（比如文字颜色）和文本输入事件的监听等。

而RCTUITextView和RCTUITextField则分别继承于UITextView和UITextField，内部重写了位置、内边距等与布局相关的属性。

2. Android

Android的ReactEditText继承于原生输入组件EditText，并且包含了InputMethodManager，用于软键盘的控制，TextInput组件具体的Android原生类如图3.19所示。

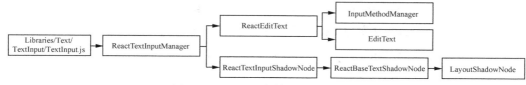

图3.19　TextInput组件Android原生类

ReactTextInputManager：TextInput组件在Android端对应的组件管理类，它负责接收TextInput传递到原生端的所有参数，并将交互相关的参数设置交由ReactEditText类，布局相关的参数设置则交由ReactTextInputShadowNode类。

ReactEditText：TextInput组件在Android端实际对应的原生组件，它基于原生基础输入框组件EditText和原生软键盘管理类InputMethodManager进行了封装，处理了输入框组件的获取焦点回调、输入状态回调、多行模式下的手势处理、软键盘类型，以及显示与隐藏等功能。

LayoutShadowNode：ReactTextInputManager中接收的布局相关的参数被设置给了LayoutShadowNode，LayoutShadowNode通过Yoga开源框架实现了Android端的Flex布局。

ReactBaseTextShadowNode：ReactTextInputManager中接收的文字格式相关的参数被设置给了React-BaseTextShadowNode，它是LayoutShadowNode的子类，实现了文字颜色等字体相关设置，以及自动换行、单行省略等格式功能。

ReactTextInputShadowNode：ReactTextInputManager中接收的光标、占位字符串、文字内容设置等输入框组件特有的相关参数被设置给了ReactTextInputShadowNode。它是ReactBaseTextShadowNode的子类，通过封装原生基础输入框组件EditText的方法实现了这些功能。

在HTML中，用单独的标签<textarea>处理多行文本输入，而在React Native中并没有独立的组件，需要使用TextInput的multiline属性来输入多行文字。而对于设置初始高度和可展示行数，iOS和Android的表现则略有不同。

（1）若只使用style的height设置高度，iOS和Android上的行为基本一致，除了由于字体默认高度不同导致能显示的行数略有偏差。TextInput内文字超出部分会形成一个内部滚动，可以通过光标来移动。

```
// 下图中第二个TextInput
<TextInput
  multiline
  textAlignVertical={'top'}
  style={{ width: 300, height: 40, borderWidth: 1, padding: 0 }}
  value={this.state.input}
  onChangeText={text => this.setState({ input: text })}
/>
```

（2）若只使用属性numberOfLines（仅在Android上生效）而不设置height值，则可初始化出一个匹配行数高度的TextInput，且当文字超出指定行数时，TextInput的高度会自行增长。iOS中虽然也具备自行增长这个特性，但缺失由numberOfLines指定的默认高度。

```
// 下图中第三个TextInput
<TextInput
  multiline
  numberOfLines={2}
  textAlignVertical={'top'}
  style={{ width: 300, borderWidth: 1, padding: 0 }}
  value={this.state.input}
  onChangeText={text => this.setState({ input: text })}
/>
```

实际效果如图3.20所示。

图3.20　TextInput组件效果

因此，为了更准确地控制输入组件位置，通常会在TextInput外部再包裹一个View，在这个View上定义边距、高度等样式，去除TextInput默认的margin和paddding，使用户输入的文字紧贴TextInput组件本身。

3.4 软键盘

在使用TextInput的时候，组件获取或失去焦点会自动开启或关闭软键盘，无须开发者手动控制。但在一些特殊场景或操作下，我们也希望能手动控制软键盘。对此，React Native提供了对应的模块及API，包括可以监听软键盘的相关事件。

3.4.1 Keyboard

Keyboard模块可以用于监听软键盘的相关事件及手动关闭软键盘，具体的使用方式如下：

```
import { Keyboard } from 'react-native';
……
componentDidMount () {
  this.keyboardDidShowListener = Keyboard.addListener('keyboardDidShow', this.keyboardDidShow);
  this.keyboardDidHideListener = Keyboard.addListener('keyboardDidHide', this.keyboardDidHide);
}

// 组件销毁前移除监听事件
componentWillUnmount() {
  this.keyboardDidShowListener.remove();
  this.keyboardDidHideListener.remove();
}

keyboardDidShow() {
  console.log('keyboardDidShow');
}

keyboardDidHide() {
  console.log('keyboardDidHide');
}
```

Keyboard的事件主要包括：keyboardWillShow、keyboardDidShow、keyboardWillHide、keyboardDidHide、keyboardWillChangeFrame、keyboardDidChangeFrame。我们可以给Keyboard绑定所有事件，并进行获取、失去焦点的操作，以观察这些事件的先后顺序，具体流程如下。

（1）获取焦点。

iOS: keyboardWillChangeFrame->keyboardWillShow->keyboardDidChangeFrame->keyboardDidShow

Android: keyboardDidShow

(2)失去焦点。

iOS：keyboardWillChangeFrame->keyboardWillHide->keyboardDidChangeFrame->keyboardDidHide

Android: keyboardDidHide

事件触发时，React Native会返回参数，用于描述此次行为的相关键盘属性。不过，不同平台返回的属性并不一致，下面将分别对其进行介绍。

1. iOS

每个事件均会返回包含以下结构的对象。

startCoordinates：事件初始时的键盘Frame属性，包括宽/高、位置等基本属性。

endCoordinates：同startCoordinates，返回的是事件结束后的键盘Frame属性。

duration：键盘动画时长。

easing：键盘动画缓动函数。

isEventFromThisApp：键盘是否属于当前App。

2. Android

相比iOS，Android Keyboard事件传递的参数较少，仅在keyboardDidShow事件中返回一个endCoordinates，keyboardDidHide事件的参数则为null。

除了监听事件之外，Keyboard还提供了手动关闭软键盘的方法——dismiss。在TextInput的JavaScript文件同级目录下的TextInputState.js文件中，可以看到dismiss具体的实现方式：

```
function blurTextInput(textFieldID: ?number) {
  if (currentlyFocusedID === textFieldID && textFieldID !== null) {
    currentlyFocusedID = null;
    if (Platform.OS === 'ios') {
      UIManager.blur(textFieldID);
    } else if (Platform.OS === 'android') {
      UIManager.dispatchViewManagerCommand(
        textFieldID,
        UIManager.getViewManagerConfig('AndroidTextInput').Commands
          .blurTextInput,
        null,
      );
    }
  }
}
```

每个React Native组件都会有唯一标识ID（在大部分情况下被命名为reactTag），UIManager为各平台实现的视图管理器的类，用于管理原生视图。上述代码在iOS中的作用是通过ReactTag找到当前原生的Responder，并且取消这个Responder的第一响应者（FirstResponder）的位置；在Android上则是利用唯一标识找到此刻唤起软键盘的ReactTextEdit，直接执行使其失去焦点的方法，从而关闭软键盘。

3.4.2　KeyboardAvoidingView

在用户单击TextInput获取焦点时，都希望唤起的键盘不会遮挡输入区域。在iOS中，通常需要监听键盘开启、关闭通知，手动控制UI视图；而在Android上，则需要在AndroidManifest.xml中设置当前Activity的android:windowSoftInputMode属性为adjustPan或adjustResize，但在一些特殊场景中（例如处于沉浸式状态栏或透明状态栏情况下）会有失效的情况。

React Native给出了一个纯粹的View层面的解决方案，核心也是使用上述的可监听事件。在Libraries/Components/Keyboard/KeyboardAvoidingView.js中我们可以看到：

```
// 计算键盘高度
_relativeKeyboardHeight(keyboardFrame): number {
  const frame = this._frame; // this._frame为最外层View的layout
  if (!frame || !keyboardFrame) {
    return 0;
  }

  const keyboardY = keyboardFrame.screenY - this.props.keyboardVerticalOffset;

  // Calculate the displacement needed for the view such that it
  // no longer overlaps with the keyboard
  return Math.max(frame.y + frame.height - keyboardY, 0);
}

_onKeyboardChange = (event: ?KeyboardEvent) => {
  if (event == null) {
    this.setState({bottom: 0});
    return;
  }

  const {duration, easing, endCoordinates} = event;
  const height = this._relativeKeyboardHeight(endCoordinates);

  if (this.state.bottom === height) {
    return;
  }

  if (duration && easing) {
```

```
    LayoutAnimation.configureNext({
      // We have to pass the duration equal to minimal accepted duration defined
here: RCTLayoutAnimation.m
      duration: duration > 10 ? duration : 10,
      update: {
        duration: duration > 10 ? duration : 10,
        type: LayoutAnimation.Types[easing] || 'keyboard',
      },
    });
  }
  this.setState({bottom: height});
};

componentDidMount(): void {
  if (Platform.OS === 'ios') {
    this._subscriptions = [
      Keyboard.addListener('keyboardWillChangeFrame', this._onKeyboardChange),
    ];
  } else {
    this._subscriptions = [
      Keyboard.addListener('keyboardDidHide', this._onKeyboardChange),
      Keyboard.addListener('keyboardDidShow', this._onKeyboardChange),
    ];
  }
}
```

通过监听Keyboard显示、隐藏或改变大小的事件，修改最外层的View的位置，以确保软键盘调出的时候获取焦点的TextInput不被遮挡。

3.5 本章小结

React Native提供了近似于Web开发方式的Text组件，但在布局和样式方面对比HTML与CSS规范仍表现出了一定的不同之处，所以开发者在使用的过程中不要想当然地照搬原先Web开发时的经验，适当减少Text组件的嵌套使用，避免出现不可预知的问题。

第 4 章 事件响应机制

对于前端开发者而言,熟悉平台的事件响应机制是成为一个合格开发者的基本条件。例如:浏览器中的冒泡与捕获、iOS 和 Android 的事件分发与消费基本都定义了视图 UI 组件该如何准确地响应用户的操作,包括事件的触发时机、事件的相关属性,以及区域重叠且绑定了相同事件的视图对事件的响应顺序,并且它们是否能够互相影响。这些基本上就是日常开发中需要弄明白的事件响应问题。

React Native 作为一个跨平台的开发框架,它提供的事件机制又是如何的呢?面对不同平台之间的差异,它又是如何进行抹平的?本章就和大家来分析一下 React Native 中事件响应机制的用法及实现。

4.1 触摸事件

在移动端 App(应用或应用程序)中,最基本、最常见的用户交互就是"触摸",也就是 Touch、Press 这类事件(不同平台惯用名称不一样)。React Native 中的部分组件直接就可以实现这类事件,例如组件 Button:

```
<Button
  onPress={() => {
    Alert.alert('Button onPress');
  }}
  title="按钮"
/>
```

上述代码会在 iOS 中生成一个默认蓝色标签状按钮(Android 则默认为蓝色圆角矩形白字按钮),onPress 则监听了用户的完整点击行为,从手指触摸屏幕到离开屏幕,算为一次 Press 行为。不过这个组件的使用率相当低,且本身也不支持 style 属性修改样式,并不能满足实际的使用需求。

除了 Button 组件外,Text 组件也能实现 Press 的相关事件,可以直接使用,例如:

```
<Text
  onPress={() => {
```

```
      Alert.alert('Text pressed');
    }}
  >
    Text 组件
  </Text>
```

在Text组件中，除onPress之外，还有几个相关参数也与Press交互相关。

onLongPress：用户长按此组件时触发，此时手指尚未离开屏幕；onLongPress触发后，手指再离开屏幕，不会触发onPress。

suppressHighlighting：默认长按会有一个灰色高亮，将该属性设置为true，去除高亮。

pressRetentionOffset：参数格式为pressRetentionOffset={{ top: 30, left: 30, right: 30, bottom: 30 }}，作用为当已触发当前Text时（suppressHighlighting若不为true，此时会有灰色高亮），手指移动出Text区域，但不超过pressRetentionOffset设定的范围，仍算是激活状态（灰色高亮不消失）。

不过在日常开发中，除了触摸这种类似于点击的交互，通常还会遇到拖曳、左右滑动等手势类操作，显然只靠监听onPress事件是不够的。因此在基础视图View组件中，React Native实现了一套类似于浏览器事件命名方式的Touch事件，例如：

```
<View
  onTouchStart={e => {
    console.log('View TouchStart', e.nativeEvent);
  }}
  onTouchMove={e => {
    console.log('View TouchMove', e.nativeEvent)
  }}
  onTouchEnd={e => {
    console.log('View TouchEnd', e.nativeEvent)
  }}
>
  <Text style={{ fontSize: 30 }}>View事件</Text>
</View>
```

触发时机基本与浏览器中的事件机制一致：手指接触到屏幕触发onTouchStart，移动手指时触发onTouchMove，手指离开屏幕触发onTouchEnd。同时，Touch事件会返回相关的参数，与位置相关的信息则主要包含在nativeEvent中，具体属性如下所示。

changedTouches：自上一次事件以来的触摸事件数组。

identifier：触摸事件ID。

locationX：触摸事件相对触发元素位置的X坐标。

locationY：触摸事件相对触发元素位置的Y坐标。

pageX：触摸事件相对根元素位置的X坐标。
pageY：触摸事件相对根元素位置的Y坐标。
target：接收触摸事件的元素ID。
timestamp：触摸事件的时间标记，用来计算速度。
touches：屏幕上所有当前触摸事件的数组。

其中，根元素指React Native视图的最外层组件，通常为整个屏幕的大小，除非当前的React Native视图嵌套在原生页面中，且只渲染屏幕的一部分。

和浏览器机制不同的是，只要触发了View的onTouchStart，即使手指已经移出View的范围，依旧能触发onTouchMove事件。这虽然有点违背之前的习惯，但对于开发拖曳效果来说倒是方便了，不用担心手指移动过快导致脱离View的触发范围。

4.2 Touch组件

除了上一节描述的UI组件和事件触发机制外，React Native中最常用的处理触摸事件的方法是使用内置的一系列Touch组件，主要包括TouchableHighlight、TouchableNativeFeedback、TouchableOpacity和TouchableWithoutFeedback。它们之间的差异如下。

TouchableWithoutFeedback：嵌套组件并提供触摸事件，但没有任何视觉上的反馈。

TouchableOpacity：降低嵌套组件的不透明度，以给人视觉上的触摸感。

TouchableHighlight：除降低嵌套组件的不透明度之外，还显示了一个由属性underlayColor设置的颜色图层，默认为黑色。

TouchableNativeFeedback：仅支持Android，直接使用Android原生的水波纹效果。

具体的使用方式如下：

```
<TouchableWithoutFeedback onPress={() => {}}>
  <View style={{ borderWidth: 1 }}>
    <Text style={{ width: 300, fontSize: 34 }}>TouchableWithoutFeedback</Text>
  </View>
</TouchableWithoutFeedback>

<TouchableHighlight onPress={() => {}}>
  <View style={{ borderWidth: 1 }}>
    <Text style={{ width: 300, fontSize: 34 }}>TouchableHighlight</Text>
  </View>
</TouchableHighlight>

<TouchableOpacity onPress={() => {}}>
```

```
  <View style={{ borderWidth: 1 }}>
    <Text style={{ width: 300, fontSize: 34 }}>TouchableOpacity</Text>
  </View>
</TouchableOpacity>

<TouchableNativeFeedback onPress={() => {}}>
  <View style={{ borderWidth: 1 }}>
    <Text style={{ width: 300, fontSize: 34 }}>TouchableNativeFeedback</Text>
  </View>
</TouchableNativeFeedback>
```

除onPress之外，Touch系列组件还提供了onPressIn和onPressOut来监听类似于onTouchStart和onTouchEnd触发时机的事件，并且可以通过设置delayPressIn和delayPressOut的值控制手势结束到最终触发事件的延迟时间。例如：

```
<TouchableWithoutFeedback
  onPressIn={() => {
    this.start = new Date().getTime();
    console.log('onPressIn', this.start)
  }}
  onPressOut={() => {
    console.log('onPressOut', new Date().getTime() - this.start)
  }}
  onPress={() => {
    console.log('onPress', new Date().getTime() - this.start)
  }}
  onLongPress={() => {
    console.log('onLongPress', new Date().getTime() - this.start)
  }}
>
  <View style={{ borderWidth: 1 }}>
    <Text style={{ width: 300, fontSize: 34 }}>TouchableWithoutFeedback</Text>
  </View>
</TouchableWithoutFeedback>
```

分别测试长按和触摸两种交互形式，输出的事件顺序及时间如图4.1所示。

长按：onPressIn=>onLongPress=>onPressOut；onLongPress触发的时机基本在500ms左右，onPressOut则为手指离开屏幕的时间，且完全取决于用户的行为。

触摸：onPressIn=>onPressOut=>onPress；若不设置onLongPress，手指离开屏幕后即会触发onPress；若设置了onLongPress，则需要手指接触屏幕的时间不超过500ms才能触发onPress，且onPress事件在

onPressOut 事件之后执行。

图 4.1　Press 触发时机（1）

下面测试一次加上 delayPressIn 和 delayPressOut 后有什么效果，例如：

```
<TouchableWithoutFeedback
  delayPressIn={1000}
  delayPressOut={1000}
  onPressIn={() => {
    this.start = new Date().getTime();
    console.log('onPressIn', this.start)
  }}
  onPressOut={() => {
    console.log('onPressOut', new Date().getTime() - this.start)
  }}
  onPress={() => {
    console.log('onPress', new Date().getTime() - this.start)
  }}
>
  <View style={{ borderWidth: 1 }}>
    <Text style={{ width: 300, fontSize: 34 }}>delayPressIn/Out</Text>
  </View>
</TouchableWithoutFeedback>
```

实际效果如图 4.2 所示。

图 4.2　Press 触发时机（2）

另外，具体的效果还分为以下两种情况。

（1）如果手指从接触到离开屏幕的时间大于delayPressIn，那刚接触屏幕时并不直接触发onPressIn，会延迟到delayPressIn设置的值之后再触发；onPressOut也是同理，所以会看到onPressOut在onPress之后触发。

（2）如果手指从接触到离开屏幕的时间小于delayPressIn，那为了事件顺序的合理性，onPressIn并不会真的延迟到delayPressIn设置的值之后触发，而会在手指离开屏幕后依次触发onPressIn、onPress或onPressOut（依据是否设置了delayPressOut）。

对于操作区域，Touch系列组件提供了hitSlop属性来扩大点击区域，但这个范围不会超过父组件边界，也不会影响原先和本组件层叠的组件。pressRetentionOffset属性则与Text中的一致，表示在当前视图不能滚动的前提下，当手指离开Touch的组件范围，但在pressRetentionOffset指定区域内，仍属于激活状态。例如：

```
<TouchableOpacity
  hitSlop={{ top: 50, bottom: 50, left: 50, right: 50 }}>
  <View style={{ width: 50, height: 50, borderWidth: 1 }}>
    <Text>hitSlop</Text>
  </View>
</TouchableOpacity>
<TouchableOpacity
  pressRetentionOffset={{ top: 50, bottom: 50, left: 50, right: 50 }}
>
  <View style={{ width: 50, height: 50, borderWidth: 1 }}>
    <Text>pressRetentionOffset</Text>
  </View>
</TouchableOpacity>
```

两者最主要的区别在于，hitSlop扩大了整个触摸范围，pressRetentionOffset仅影响触摸激活的有效范围。

4.3 手势响应系统

手势响应系统可以分为两部分来说明：手势识别和响应系统。手势识别不言而喻，移动端的交互方式比PC Web要复杂许多，那React Native提供了怎样的方式来识别用户操作的手势，开发者又该如何使用呢？响应系统则对应的是此次交互行为该触发哪个事件，特别是组件存在重叠的场景下，事件的触发顺序如何，是否可被拦截？带着这些问题，本节将具体分析React Native的手势响应系统。

4.3.1 响应者生命周期

在移动端，iOS、Android和Web平台在对事件响应的处理方面或多或少存在差异，为了能达到统一，React Native采取了三者较为通用的部分来形成自己的一套响应系统，也抛弃了各平台中一些较为独特的策略。我们可以用以下的例子来阐述React Native的响应方式：

```
<View
  style={{ width: 100, height: 50, borderWidth: 1 }}
  onStartShouldSetResponder={evt => {
    console.log('onStartShouldSetResponder', evt);
  }}
  onMoveShouldSetResponder={evt => {
    console.log('onMoveShouldSetResponder', evt);
  }}
/>
```

View组件提供了两个参数onStartShouldSetResponder和onMoveShouldSetResponder,作用为询问当前View是否响应此次触摸事件,触发的时机分别为触摸屏幕时和移动触摸点时,一旦某次询问返回值为true,则不会再次触发函数,例如:

```
<View
  style={{ width: 100, height: 50, borderWidth: 1 }}
  onStartShouldSetResponder={evt => {
    console.log('onStartShouldSetResponder', evt);
    return true;
  }}
  onMoveShouldSetResponder={evt => {
    console.log('onMoveShouldSetResponder', evt);
  }}
/>
```

如果将上述例子中的onStartShouldSetResponder返回值设置为true,则将不会再执行onMoveShouldSetResponder;同样,如果某次onMoveShouldSetResponder返回值为true,那继续移动手指也将不会再触发onMoveShouldSetResponder。

View获取此次响应之后,会继续触发onResponderGrant、onResponderMove、onResponderEnd和onResponderRelease,时机基本与onTouchStart、onTouchMove和onTouchEnd一致,但提前于Touch系列,例如:

```
<View
  style={{ width: 100, height: 50, borderWidth: 1 }}
  onStartShouldSetResponder={evt => true}
  onResponderGrant={evt => {
    console.log('onResponderGrant')
  }}
  onResponderMove={evt => {
    console.log('onResponderMove')
  }}
```

```
  onResponderEnd={evt => {
    console.log('onResponderEnd')
  }}
  onResponderRelease={evt => {
    console.log('onResponderRelease')
  }}
  onTouchStart={e => {
    console.log('View TouchStart');
  }}
  onTouchMove={e => {
    console.log('View TouchMove')
  }}
  onTouchEnd={e => {
    console.log('View TouchEnd')
  }}
/>
```

响应顺序如图4.3所示。

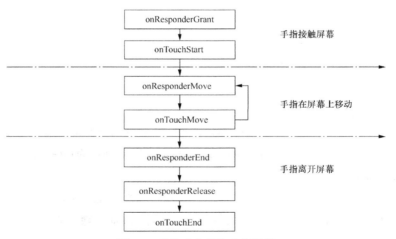

图4.3 手势响应系统响应顺序

如果存在多个组件都对某个事件进行响应的情况，流程和判断条件就相对复杂一点，下面用一个嵌套组件的例子来说明这个过程：

```
<View
  style={{ width: 100, height: 50, borderWidth: 1 }}
  onStartShouldSetResponder={evt => {
    console.log('onStartShouldSetResponder')
    return true;
```

```jsx
    }}
    onMoveShouldSetResponder={evt => {
      console.log('onMoveShouldSetResponder');
      return true;
    }}
    onResponderGrant={evt => {
      console.log('onResponderGrant')
    }}
    onResponderMove={evt => {
      console.log('onResponderMove')
    }}
    onResponderEnd={evt => {
      console.log('onResponderEnd')
    }}
    onResponderRelease={evt => {
      console.log('onResponderRelease')
    }}
>
    <View
      style={{ width: 80, height: 40, borderWidth: 1 }}
      onStartShouldSetResponder={evt => {
        console.log('onStartShouldSetResponder', 'child')
        return true;
      }}
      onMoveShouldSetResponder={evt => {
        console.log('onMoveShouldSetResponder', 'child');
        return true;
      }}
      onResponderGrant={evt => {
        console.log('onResponderGrant', 'child')
      }}
      onResponderMove={evt => {
        console.log('onResponderMove', 'child')
      }}
      onResponderEnd={evt => {
        console.log('onResponderEnd')
      }}
      onResponderRelease={evt => {
        console.log('onResponderRelease', 'child')
      }}
    />
</View>
```

以这个例子为基础，下面修改父子组件的 onStartShouldSetResponder 和 onMoveShouldSetResponder 返回值来观察整个响应过程，具体的响应顺序如图4.4所示。

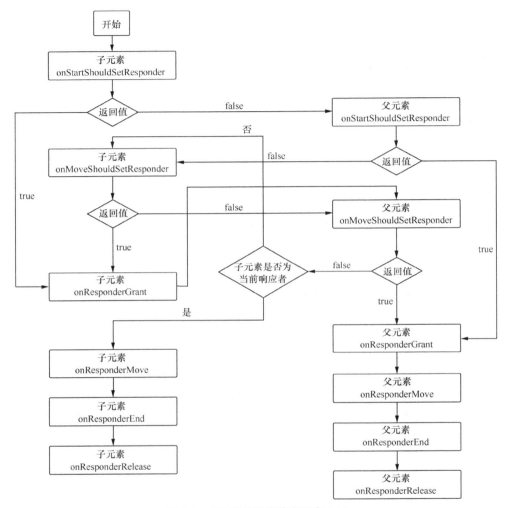

图4.4 多个手势系统响应顺序（1）

子组件会通过 onStartShouldSetResponder 首先申请成为响应者，而父组件如果一开始未成为响应者，则会在移动的过程中通过 onMoveShouldSetResponder 不停地发起询问，并且一旦成功（即父组件 onMoveShouldSetResponder 返回值为true），则子组件会失去响应者的能力，并将不再触发自己对应的 Responder 事件。

那在整个过程中，子组件是否可以拒绝父组件的响应者申请，或父组件是否能够拦截子组件的申请，从一开始就自己申请成为响应者？对此，React Native 的响应者机制提供了 onResponderTermination-Request 和 onStartShouldSetResponderCapture（onMoveShouldSetResponderCapture）事件来改变这个流程。

onResponderTerminationRequest：常作用于子组件，触发时机为有其他组件申请成为响应者时，当前组件是否允许？若返回值为true，则让出响应者，并触发onResponderTerminate事件。

onStartShouldSetResponderCapture(onMoveShouldSetResponderCapture)：若返回值为true，则当前组件会阻止子组件成为响应者。

加上这两个流程，上述流程被补充成（虚线为新加节点）图4.5。

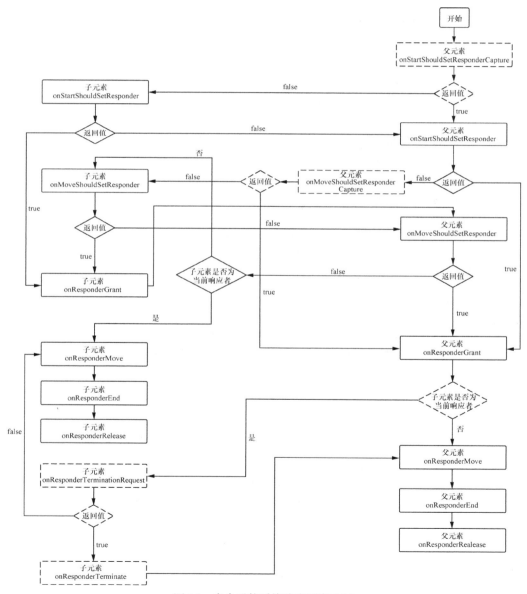

图4.5 多个手势系统响应顺序（2）

需要注意的是，onTouchStart、onTouchMove和onTouchEnd这3个事件并不在响应者体系中，它们既不能被中断，也不能被捕获，始终都会从嵌套最深的组件开始向外冒泡触发。

4.3.2 PanResponder

PanResponder是React Native中提供的一套手势识别系统，它可以将单点触摸解析成更具体的手势，也可以识别简单的多点触摸。PanResponder是对responder进行了一层分装，所以在执行流程上与前边介绍的响应者生命周期是一致的，只不过命名方式略有不同，具体用法如下。

```
export default class App extends Component {
  componentWillMount() {
    this._panResponder = PanResponder.create({
      onStartShouldSetPanResponder: (evt, gestureState) => {
        console.log('onStartShouldSetPanResponder', gestureState);
        return true;
      },
      onPanResponderGrant: (evt, gestureState) => {
        console.log('onPanResponderGrant', gestureState)
      },
      onPanResponderMove: (evt, gestureState) => {
        console.log('onPanResponderMove', gestureState)
      },
      onPanResponderEnd: (evt, gestureState) => {
        console.log('onPanResponderEnd', gestureState)
      },
      onPanResponderRelease: (evt, gestureState) => {
        console.log('onPanResponderRelease', gestureState)
      },
    });
  }

  render() {
    return (
      <View style={{ flex: 1, justifyContent: 'center', alignItems: 'center' }}>
        <View
          style={{ width: 100, height: 50, borderWidth: 1 }}
          {...this._panResponder.panHandlers}
        />
      </View>
    )
  }
}
```

单击View组件区域，可以得到图4.6所示结果。

```
onStartShouldSetPanResponder ▶ {stateID: 0.7064043586071198, moveX: 0, moveY: 0, x0: 0, y0: 0, …}      index.js:9
onPanResponderGrant ▶ {stateID: 0.7064043586071198, moveX: 0, moveY: 0, x0: 185, y0: 450.5, …}          index.js:13
onPanResponderStart ▶ {stateID: 0.7064043586071198, moveX: 0, moveY: 0, x0: 185, y0: 450.5, …}          index.js:16
onPanResponderMove ▶ {stateID: 0.7064043586071198, moveX: 185, moveY: 450.5, x0: 185, y0: 450.5, …}     index.js:19
onPanResponderMove ▶ {stateID: 0.7064043586071198, moveX: 185, moveY: 450.5, x0: 185, y0: 450.5, …}     index.js:19
onPanResponderMove ▶ {stateID: 0.7064043586071198, moveX: 185, moveY: 450.5, x0: 185, y0: 450.5, …}     index.js:19
onPanResponderEnd ▶ {stateID: 0.7064043586071198, moveX: 185, moveY: 450.5, x0: 185, y0: 450.5, …}      index.js:22
onPanResponderRelease ▶ {stateID: 0.7064043586071198, moveX: 185, moveY: 450.5, x0: 185, y0: 450.5, …}  index.js:25
```

图4.6　PanResponder触发顺序

与Responder不同的是，PanResponder需要通过调用create的方式进行创建，并将返回实例的panHandlers作为参数传给View。PanResponder.create涉及的参数主要包括以下这些。

onStartShouldSetPanResponder: (e, gestureState) => {}：触摸屏幕时询问是否成为响应者。

onStartShouldSetPanResponderCapture: (e, gestureState) => {}：触摸屏幕时询问是否捕获成为响应者（不向子元素传递）。

onMoveShouldSetPanResponder: (e, gestureState) => {}：移动时询问是否成为响应者。

onMoveShouldSetPanResponderCapture: (e, gestureState) => {}：移动时询问是否捕获成为响应者（不向子元素传递）。

onPanResponderGrant: (e, gestureState) => {}：授权成为响应者。

onPanResponderStart: (e, gestureState) => {}：开始响应。

onPanResponderMove: (e, gestureState) => {}：移动时响应。

onPanResponderEnd: (e, gestureState) => {}：响应结束。

onPanResponderRelease: (e, gestureState) => {}：响应释放。

onPanResponderTerminationRequest: (e, gestureState) => {}：有其他组件想要成为响应者时，询问当前响应者是否同意。

onPanResponderTerminate: (e, gestureState) => {}：当前响应被中断。

onPanResponderReject: (e, gestureState) => {}：当前响应被拒绝。

onShouldBlockNativeResponder: (e, gestureState) => {}：是否阻止原生响应。

另外，PanResponder的事件比Responder多了一个gestureState的参数，主要包括以下属性。

stateID：触摸状态ID。在屏幕上有至少一个触摸点的情况下，ID一直有效。

moveX：最近一次移动时的屏幕横坐标。

moveY：最近一次移动时的屏幕纵坐标。

x0：响应者产生时的屏幕横坐标。

y0：响应者产生时的屏幕纵坐标。

dx：从触摸操作开始时的累计横向距离。

dy：从触摸操作开始时的累计纵向距离。

vx：当前的横向移动速度。

vy：当前的纵向移动速度。

numberActiveTouches：当前屏幕上的有效触摸点数。

我们可以用PanResponder来实现简单的拖曳移动效果，下面讲解它的用法。

```
export default class App extends Component {
  state = {
    x: 0,
    y: 0
  };

  componentWillMount() {
    this.startX = 0;
    this.startY = 0;
    this._dragResponder = PanResponder.create({
      onStartShouldSetPanResponder: (evt, gestureState) => true,
      onPanResponderMove: (evt, gestureState) => {
        this.setState({
          x: this.startX + gestureState.dx,
          y: this.startY + gestureState.dy
        })
      },
      onPanResponderEnd: (evt, gestureState) => {
        this.startX += gestureState.dx;
        this.startY += gestureState.dy;
      }
    });
  }

  render() {
    const { x, y } = this.state;
    return (
      <View style={{ flex: 1, justifyContent: 'center', alignItems: 'center' }}>
        <View
          style={{
            width: 100,
            height: 50,
            borderWidth: 1,
```

```
              transform: [
                { translateX: x },
                { translateY: y },
              ]
            }}
            {...this._dragResponder.panHandlers}
          />
        </View>
      )
    }
  }
```

这样就可以实现拖曳的效果。当然,如果你熟悉React的话,会发现我们在onPanResponderMove中频繁调用了setState方法,这显然不是一个高效的方式,会导致组件过多地触发render方法。那有没有什么可以优化的方法呢?

4.4 原生事件机制

React Native 提供的事件机制也分为捕获和冒泡(原生端也叫分发和传递)两部分,捕获指从最外层视图一层层向内传递,直至具体触发的视图;冒泡则指从事件直接触发的视图向外传递,直至最外层视图。下面结合 Responder 来分析,基本流程如图 4.7 所示。

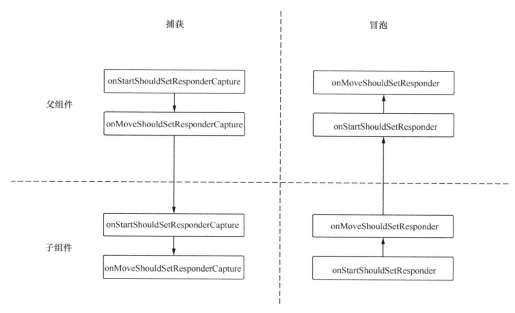

图 4.7　PanResponder 事件机制

与JavaScript开发者熟悉的浏览器机制——捕获与冒泡不同的是，React Native的冒泡不能被阻止。父组件如果获取不到响应者权限，会在事件触发过程中不断询问，子组件只能通过设置onResponderTerminationRequest返回值为true进行阻止。

React Native的事件机制本质上还是依赖于原生端原有的事件机制，封装后提供一套统一的流程，以抹去平台上的差异。为了更深入地理解这套机制，使用户更好地处理复杂的事件交互，下面介绍原生平台的事件机制，以及最终React Native如何将原生事件分发给JavaScript端的绑定函数。

4.4.1 iOS事件机制

iOS中的事件机制会涉及以下几个原生类。

UIResponder：接受Touch相关的事件，例如touchesBegan、touchesMoved、touchesEnded和touchesCancelled，当手指触摸屏幕时会首先调用touchesBegan。

UIGestureRecognizer：手势识别类，优先级比UIResponder高，当UIResponder收到touchesBegan后便开始等待UIGestureRecognizer开始识别，识别成功会放弃UIResponder的事件传递。

UIControl：优先级最高，如果视图中添加了继承于UIControl的组件（例如UIButton）并触发，将无视视图中可能存在的UIResponder和UIGestureRecognizer的事件。

在React Native中只使用了UIResponder和UIGestureRecognizer这两种方式。从第2章我们知道View组件在iOS中会被转化成RCTRootView，RCTRootView继承于iOS的UIView，再向上追溯的话会发现UIView本身就继承于UIResponder，所以View组件本身就会遵循UIResponder提供的事件机制。同时，RCTRootView中也包含了一个继承于UIGestureRecognizer的RCTTouchHandler生成的实例，相当于是使用了这一手势识别的特性。图4.8展示了这几个类的关系（图中只展示了涉及事件相关的类及属性）。

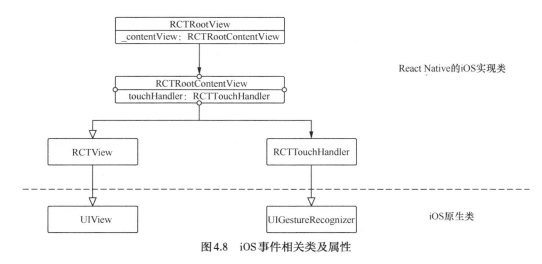

图4.8 iOS事件相关类及属性

下面以TouchWithoutFeedback组件为例，描述一下事件触发从iOS到JavaScript的流程，如图4.9所示。

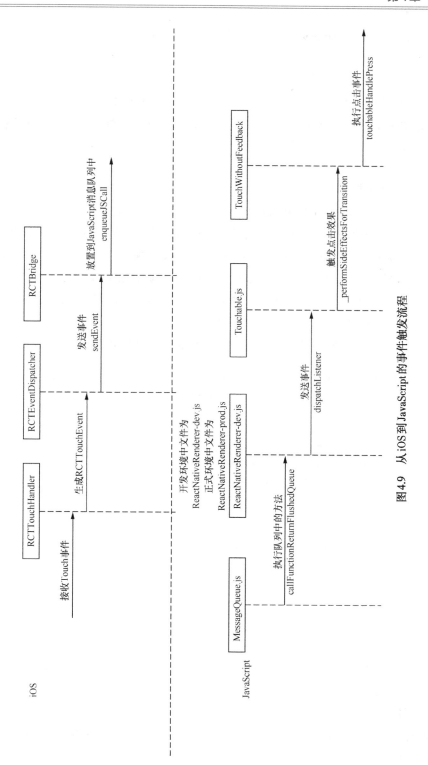

图 4.9 从 iOS 到 JavaScript 的事件触发流程

由图4.9可见，React Native在原生Event对象的基础上封装了一层RCTTouchEvent，然后通过RCTEventDispatcher将消息发送给RCTBridge。RCTEventDispatcher会单独使用JavaScript线程队列，以维护线程安全。RCTBridge会将所有需要JavaScript处理的消息统一起来，发送给JavaScript的消息队列MessageQueue.js来处理，最终交于ReactNativeRender对象，便于其根据触发的模块调用对应的函数。Touchable.js是TouchWithoutFeeback等所有组件所使用的基础对象，处理了包括接收事件和判断Responder状态等通用逻辑，最终触发Touch组件上所绑定的JavaScript事件。

4.4.2 Android事件机制

在Android中，一切视图组件都继承自View和ViewGroup类，其中ViewGroup本质上也是继承自View类的，只不过比View增加了管理子组件的功能，而View只能作为视图树上的叶子节点。基础视图类View中包含onTouchEvent和dispatchTouchEvent两个方法来控制事件的流程。从用户触摸屏幕开始，Android中由最底层的Activity最先响应，将事件一层层传递到对应的View（此处的View是指Android原生的View类），如果View响应了该事件，则称之为事件被消费了，该View的父View将不再接收这个事件；反之，若子View未消费这个事件，则可以选择交由父View对其进行处理。ViewGroup类通过继承View类的方法也具有相同的事件机制（见图4.10），大致的代码如下。

```java
public class TouchView extends View {
    ......

    @Override
    public boolean dispatchTouchEvent(MotionEvent event) {
        Log.d("View dispatchTouchEvent" + event.getAction());
        return super.dispatchTouchEvent(event);
    }

    @Override
    public boolean onTouchEvent(MotionEvent event) {
        Log.d("View onTouchEvent" + event.getAction());
        return super.onTouchEvent(event);
    }
    ......
}
```

其中，dispatchTouchEvent在任何时候都会优先执行，并且在内部调用了onTouch和OnTouchEvent方法，它们之间存在如下关系。

当onTouch返回值为true时，dispatchTouchEvent返回值也为true，onTouchEvent方法将不再执行。

当onTouch返回值为false时，会执行onTouchEvent方法。onTouchEvent方法的返回值也就决定了dispatchTouchEvent的返回值。

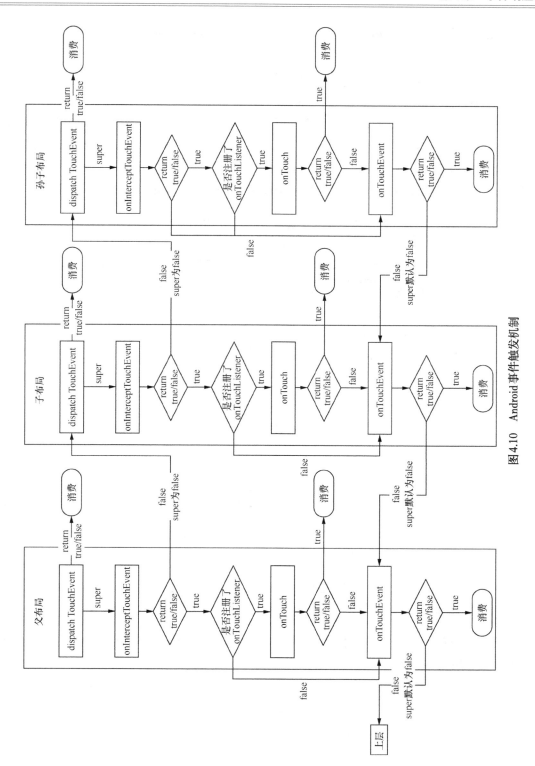

图 4.10 Android 事件触发机制

当dispatchTouchEvent返回值为true时，表示消费事件，该View后续的事件将继续响应，并不再向下分发该事件。

当dispatchTouchEvent返回false时，表示不消费事件，交由子View去处理。

React Native在Android上的View实现类ReactRootView继承于Android原生的FrameLayout及ViewGroup，所以它包含了dispatchTouchEvent、onInterceptTouchEvent和onTouchEvent等事件相关函数。ReactRootView中的JSTouchDispatcher主要用于将原生的Touch事件封装成React Native自定义的Touch事件，然后调用EventDispatcher的dispatchEvent方法，最终会在React Native在Android中创建的上下文——ReactContext中利用队列的方式将自定义的Touch事件传递给JavaScript端，具体类之间的关系如图4.11所示。

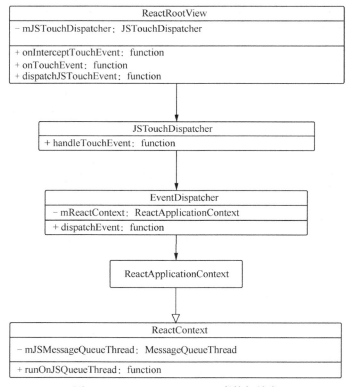

图4.11　React Native Android事件相关类

下面仍以TouchWithoutFeedback组件为例，描述事件触发从Android到JavaScript的流程，如图4.12所示。

同样由原生视图ReactRootView负责接收用户触发的Touch事件，交由JSTouchDispatcher进行二次封装，然后进入EventDispatcher中的事件队列等待发送。发送的过程中，使用了Android的Choreographer（编舞者）模式，可以在UI重绘时（间隔为16ms）调用React Native上下文（ReactApplicationContext）中的runOnJSQueueThread，在JavaScript线程中发送队列中等待的所有事件。

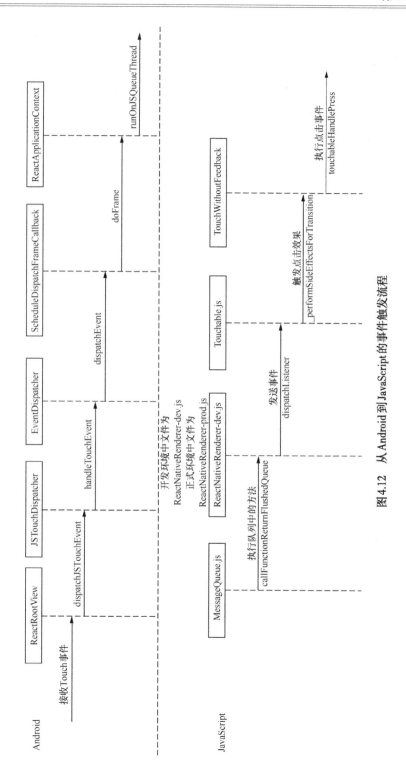

图 4.12　从 Android 到 JavaScript 的事件触发流程

4.5 本章小结

整体来讲，React Native 的事件机制与浏览器大同小异，但在处理多个组件同时响应的情况时，并没有浏览器中常用的 preventDefault 和 stopPropagation 这样的方法。从另一个角度来说，React Native 并没有控制响应事件的流转，而是从组件的角度出发去设置是否对此次事件进行响应。

第5章 媒体、文件及本地存储

近年来，随着移动互联网速度的不断提升，以及流量资费的下降，人们接受信息的方式也从文字、图片逐渐转变成了音频、视频等更多样化的载体。这些变化对前端开发者提出了新的要求和挑战，开发者需要熟练地在对应的平台或系统上使用、处理这些多媒体格式，并对其常见的性能问题有一定的优化手段和策略。React Native 作为一个跨平台的解决方案，也提供了一些基础的组件和 API 来支持这些功能，例如图片的展示和简单的本地存储。而对于类似于音频、视频，以及本地文件的操作等功能，官方并没有提供，我们也就只能求助于社区提供的插件了。本章将会具体分析 React Native 对媒体格式的支持程度及实现方式，也会介绍目前常见的一些插件方案。如果读者所在的团队本身具备一定原生开发能力的话，可以自行开发所需的原生组件或模块来提升 React Native 对媒体、文件和存储的支持。

5.1 图片组件

React Native 提供了 Image 与 ImageBackground 两个和图片相关的组件，用于显示静态资源、网络图片、本地的临时图片及磁盘上的图片，基本用法如下。

```
// 静态资源
<Image
  source={require('./img/example.png')}
/>
// 网络图片
<Image
  style={{width: 50, height: 50}}
  source={{uri: 'https://facebook.github.io/react-native/img/tiny_logo.png'}}
/>
// 本地文件
<Image
  style={{ width: 50, height: 50 }}
```

```
    source={{ uri: 'content://com.android.providers.media.documents/document/ic_
close.png' }}
  />
  // 支持base64格式
  <Image
    style={{width: 66, height: 58}}
    source={{uri: 'data:image/png;base64,iVBORw0KGgoAAAANSUhEUgAAADMAAAAzCAYAA
AA6oTAqAAAAEXRFWHRTb2Z0d2FyZQBwbmdjcnVzaEB1SfMAAABQSURBVGje7dSxCQBACARB+2/ab8BEe
QNhFi6WSYzYLYudDQYGBgYGBgYGBgYGBgZmcvDqYGBgmhivGQYGBgYGBgYGBgYGBgbmQw+P/
eMrC5UTVAAAAABJRU5ErkJggg=='}}
  />
```

需要注意的是，如果设置的图片资源为网络图片或base64数据资源，那么必须要对图片的尺寸进行手动设置。除了基本的宽/高属性，Image的style属性还提供了borderRadius（圆角）和shadowOffset（阴影）等常见的参数设置，但Android不支持设置阴影参数，而iOS在圆角和阴影两个属性中只能生效一个。所以，如果想要两个效果叠加，通常需要嵌套视图来设置样式，代码如下。

```
const shadowsStyling = {
  width: 100,
  height: 100,
  shadowColor: "#000000",
  shadowOpacity: 0.8,
  shadowRadius: 2,
  shadowOffset: {
    height: 1,
    width: 0
  }
}

<View style={shadowsStyling}>
  <Image
    style={{ borderRadius: 10 }}
    ......
  />
</View>
```

与Image组件不同的是，ImageBackground组件支持嵌套用法，允许在组件内嵌套其他的组件形式，例如：

```
<ImageBackground
  style={{ width: 50, height: 50 }}
```

```
  source={{ uri: 'https://facebook.github.io/react-native/img/tiny_logo.png'}}
>
  <View style={{ width: 30, height: 30, background: '#FFFFFF' }} />
</ImageBackground>

<ImageBackground
  style={{ width: 50, height: 50 }}
  source={{ uri: 'https://facebook.github.io/react-native/img/tiny_logo.png'}}
>
  <Text>背景为图片的文字</Text>
</ImageBackground>
```

5.1.1 Image属性及方法详解

除style及source这两个基本属性外,Image还提供了其他属性和方法来帮助开发者控制图片的展示及加载,常见的属性包括以下几种。

resizeMode:若图片的分辨率尺寸和Image组件的尺寸不匹配,需要指定图片调整尺寸的模式,一共有5种模式可供选择,默认为'cover'。

'cover':保持图片比例对图片进行缩放,直至宽和高都大于等于Image组件设置的尺寸。

'contain':保持图片比例对图片进行缩放,直至宽和高都小于等于Image组件设置的尺寸。

'stretch':不保持图片比例,拉伸图片至Image组件设置的尺寸大小。

'repeat':保持图片比例,无法填充Image组件的部分被重复的图片平铺。

'center':居中图片,若图片尺寸大于Image组件的尺寸,则将图片缩小至能够被完全显示,然后居中显示。

具体代码示例如下。

```
<View style={{ justifyContent: 'center', alignItems: 'center', flex: 1 }}>
  <Text style={{textAlign: 'center'}}>
    以下Image组件均设置为width: 50, height: 100, resizeMode设置为不同的值的效果如下:
  </Text>
  <View style={{ flexWrap: 'wrap', flexDirection: 'row' }}>
    <View style={{alignItems: 'center', margin: 5}}>
      <Text>cover</Text>
      <Image
        style={{
          height: 100,
          width: 50,
          backgroundColor: 'blue'
```

```
      }}
      resizeMode={'cover'}
      source={{uri: 'https://video512.oss-cn-beijing.aliyuncs.com/IMG_6959.jpeg'}}
    />
</View>

<View style={{alignItems: 'center', margin: 5}}>
  <Text>contain</Text>
  <Image
    style={{
      height: 100,
      width: 50,
      backgroundColor: 'blue'
    }}
    resizeMode={'contain'}
    source={{uri: 'https://video512.oss-cn-beijing.aliyuncs.com/IMG_6959.jpeg'}}
  />
</View>

<View style={{alignItems: 'center', margin: 5}}>
  <Text>stretch</Text>
  <Image
    style={{
      height: 100,
      width: 50,
      backgroundColor: 'blue'
    }}
    resizeMode={'stretch'}
    source={{uri: 'https://video512.oss-cn-beijing.aliyuncs.com/IMG_6959.jpeg'}}
  />
</View>

<View style={{alignItems: 'center', margin: 5}}>
  <Text>repeat</Text>
  <Image
    style={{
      height: 100,
      width: 50,
      backgroundColor: 'blue'
    }}
```

```
      resizeMode={'repeat'}
      source={{uri: 'https://video512.oss-cn-beijing.aliyuncs.com/IMG_6959.jpeg'}}
    />
  </View>

  <View style={{alignItems: 'center', margin: 5}}>
    <Text>center</Text>
    <Image
      style={{
        height: 100,
        width: 50,
        backgroundColor: 'blue'
      }}
      resizeMode={'center'}
      source={{uri: 'https://video512.oss-cn-beijing.aliyuncs.com/IMG_6959.jpeg'}}
    />
  </View>
 </View>
</View>
```

实际效果如图5.1所示。

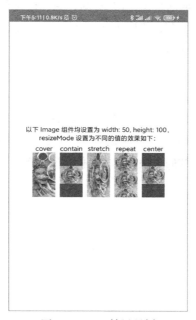

图5.1　Image效果示例

defaultSource：在图片加载过程中显示的默认占位图（Android只有在release版本中才生效）。

tintColor：将图片中的非透明像素指定成当前颜色。

在图片格式方面，如果需要在Android上显示GIF或WebP等其他格式的图片，还要在对应工程下的'android/app/build.gradle'文件中加入这些依赖：

```
dependencies {
  // 如果你的Android App支持的版本低于Android 4.0 Ice Cream Sandwich (API level 14)
  implementation 'com.facebook.fresco:animated-base-support:1.10.0'

  // 支持GIF格式的动态图片
  implementation 'com.facebook.fresco:animated-gif:1.12.0'

  // 支持WebP格式的动态图片
  implementation 'com.facebook.fresco:animated-webp:1.10.0'
  implementation 'com.facebook.fresco:webpsupport:1.10.0'

  // 支持WebP格式的非动态图片
  implementation 'com.facebook.fresco:webpsupport:1.10.0'
}
```

Image除了作为组件之外，同时也可作为功能模块使用，它提供了如下方法。

getSize/getSizeWithHeaders：获取网络图片的宽/高，这一过程包含了下载图片，理论上可以用这一方法来实现预加载，但官方并不推荐，建议还是使用专门的prefetch方法。

prefetch：预加载方法，提前将图片下载到本地缓存中。

abortPrefetch：停止预加载，且仅支持Android。

queryCache：查询图片缓存状态。

resolveAssetSource：解析本地图片，获取其路径及宽/高。

利用上述这些属性和方法，我们可以在类似轮播图的场景中提前加载所有的图片，避免用户在滑动过程中出现图片未加载完成而导致的卡顿情况，示例代码如下。

```
import React, { Component } from 'react';
import { Image } from 'react-native';

export default class SimpleSlider extends Component {

  constructor(props) {
    super(props);
    const { prefetchImgs } = props;
    if (prefetchImgs && prefetchImgs instanceof Array) {
```

```
        prefetchImgs.forEach( img => { Image.prefetch(img) } );
      }
    }
    ...
    render() {
      const { defaultSource = require('../5.1/icon.png'), prefetchImgs, index, ...
restProps } = this.props;
      return (
        <Image
          defaultSource={defaultSource}
          source={{ uri: prefetchImgs [index] }}
          { ...restProps }
        />
      );
    }
}
```

5.1.2 原生图片组件

同Text组件类似,Image组件最终也会被解析成各原生端的基础图片UI组件,而React Native通常会对其进行封装,一是兼容平台之间的差异,二是补充功能,使之能更符合Web开发者的习惯。Image组件在iOS和Android上主要的实现类分别为RCTImageView和ReactImageView,下面具体分析一下这两个类的功能及上下游的调用关系。

1. iOS

在iOS中,图片通常由原生UI类UIImage管理,UIImageView负责展示,并提供了加载网络和本地图片的能力。React Native中的RCTImageView并没有直接继承于UIImageView,而是继承于最基础的UIView,在类中自行管理了UIImage对象,具体关系如图5.2所示。

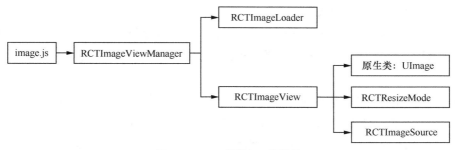

图5.2 Image组件iOS相关类

RCTImageViewManager：ReactImageView 的管理类，导出 API 供 JavaScript 端调用。

RCTImageLoader：负责加载生成 UIImage 对象，也负责管理图像中包含的信息（方向和透明通道等）、处理 resize 和管理加载进度。

RCTImageSource：管理 image 对象的 source，如 requestURL、sacl 和 size 等。

RCTResizeMode：对应 JavaScript 端设置的 resize 类型，如 cover 和 center 等。

2. Android

Android 原生系统中常用的图片 UI 组件为 ImageView，它既支持在 xml 布局文件中静态设置 drawable 对象，也可以在 Java 文件中动态设置 drawable 对象、资源 id、Bitmap 对象和 Uri 对象。但 ImageView 的功能略为单一，并不支持直接设置网络图片路径，也没有使用分级缓存的优化策略，所以 React Native 并没有直接使用它进行二次封装，而是继承了 Facebook 自己的图片加载框架 Fresco 中的 GenericDraweeView 类，来实现图片加载、展示这一系列的功能，具体关系如图 5.3 所示。

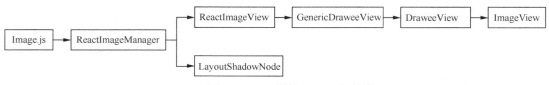

图 5.3　Image 组件 Android 相关类

ReactImageManager：主要负责从 JavaScript 层获取组件设置的参数值，如 source、defaultSource、tintColor 等，然后将参数值赋值给 ReactImageView，也负责将 ReactImageView 与 LayoutShadowNode 建立关联。

LayoutShadowNode：接收 JavaScript 层设置的布局相关的参数值，通过 Yoga 框架完成了 Flex 布局在原生层的实现。

ReactImageView：JavaScript 层 Image 组件在 Android 原生层对应的图像加载组件，完成了对 Fresco 图像加载框架的封装，可以接收 JavaScript 层设置的参数并进行配置。

GenericDraweeView：ReactImageView 的父类，封装了 Fresco 图像加载框架，是 Fresco 提供的可以在 Android 原生层直接使用的图像加载组件。

DraweeView：GenericDraweeView 的父类，Fresco 图像加载框架的具体实现类，可以将图像 drawable、圆角裁剪 drawable、占位图 drawable 和加载进度条 drawable 等多个 drawable 绘制到画布上，实现比原生 ImageView 更强大的功能。

ImageView：Android 原生的图像展示 UI 组件，DraweeView 仅仅使用了它的 padding 计算方法和绘制方法，所以 Facebook 官方打算在后续版本中用 View 替换它，作为 DraweeView 的父类。

ReactImageManager 在接收了 JavaScript 层设置的参数后，将与布局相关的参数设置给 LayoutShadowNode 以实现 Flex 布局，将与图片资源和样式相关的参数通过 ReactImageManager 设置给 ReactImageView。

也就是说，LayoutShadowNode完成了与布局相关的工作，ReactImageView完成了与图片及样式渲染相关的工作。ReactImageView因为底层采用了Fresco图片加载框架，所以可以完成网络图片的加载、静态本地资源的加载、图片的缓存逻辑等。

Fresco中核心的类是DraweeView类，它在图片渲染方面与Android原生的ImageView组件最大的不同是，它不仅将图片资源drawable对象绘制到了画布上，也将各种圆角裁剪样式、占位图资源、加载进度条动画样式等多个drawable对象进行管理和统一调整，最终绘制到画布上，实现了更多、更复杂的功能。

5.1.3 高性能图片组件：react-native-fast-image

虽然Image组件实现了诸多的功能，也满足了日常的业务场景，但在性能这一方面没有做太多的优化，在需要同时展示大量图片的场景下，经常会出现闪烁或加载失败的现象。为了解决这个问题，社区提供了基于iOS图片加载框架SDWebImage和Android图片加载框架Glide的react-native-fast-image插件，并逐渐成为目前React Native项目中最为广泛使用的图片加载框架之一。

与Image组件相比，react-native-fast-image主要提供了以下几种额外功能。

（1）更具体、多样的图片缓存策略。
（2）支持设置加载优先级，包括low、normal和high 3种方式。
（3）直接支持GIF，不需要在Android再额外引用其他依赖。

在使用方式上，react-native-fast-image做到了尽量与官方Image组件API保持一致，且大部分较新的版本已要求项目中的React Native升级到0.60.0以上，具体代码如下。

```
import React, { Component } from 'react';
import { View } from 'react-native';
import FastImage from 'react-native-fast-image';

export default class App extends Component{

  render() {
    return (
      <View style={{flex: 1, justifyContent: 'center', alignItems: 'center'}}>
        <FastImage
          style={{ width: 200, height: 200 }}
          source={{
            uri: 'https://unsplash.it/400/400?image=1',
            headers: { Authorization: 'someAuthToken' },
            priority: FastImage.priority.normal,
          }}
```

```
            resizeMode={ FastImage.resizeMode.contain }
        />
      </View>
    );
  }
};
```

在react-native-fast-image提供的source属性中，还可以设置具体的缓存策略，例如source.cache可以设置以下3种不同的缓存枚举类型。

FastImage.cacheControl.immutable：url变化后才更新文件。

FastImage.cacheControl.web：使用Headers，并遵循常规的缓存策略。

FastImage.cacheControl.cacheOnly：仅从缓存中读取文件。

与Image的缓存策略对比，尽管Fresco是一个功能非常强大的图片加载框架，支持多种缓存策略的配置，但是React Native在封装Image组件时将Fresco配置成了一种尽量不占用内存的缓存策略，如图5.4所示。

可以看出Image组件的缓存策略是将网络图片总是先下载至硬盘中，然后根据页面中要渲染哪些页面，再将其从硬盘读取至内存，完成渲染。页面关闭后，清空内存中保留的图片信息，但不清除硬盘中缓存的图片。再次进入页面时，依然先检查硬盘中缓存的图片，然后将图片读取至内存，最后完成渲染。

这样的策略极大地节省了内存的占用，但更加耗费硬盘的存储空间，并且从硬盘每次读取到内存的读写开销也占了比较长的时间，因此如果使用Image组件，每次进入新的页面时都会出现图片闪烁的现象。目前Image并未提供方法来清除本地硬盘上缓存的图片文件，如果开发者需

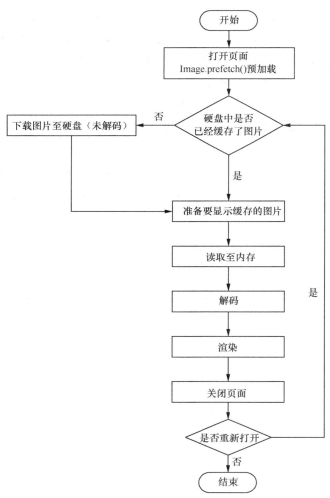

图5.4 图片缓存策略（1）

要保持应用在本地磁盘的存储空间占用情况，还需要自己查看这些图片缓存在本地硬盘的存储路径，然后手动编写清除缓存的方法来清除这些数据。

虽然react-native-fast-image在iOS采用了开源图片加载框架SDWebImage，在Android采用了开源图片加载框架Glide进行封装，但是它在两个原生端使用的缓存策略是一致的。图5.5展示了react-native-fast-image使用的图片缓存策略。

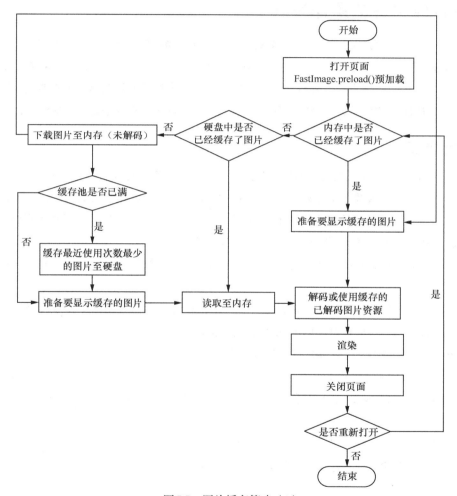

图5.5　图片缓存策略（2）

概括地讲，react-native-fast-image采用了先内存后硬盘的缓存策略。当打开新的页面加载网络图片时，会优先将图片下载至内存中缓存，如果内存中的图片大小大于预设的缓存池容量，才会将图片以"最近一段时间内使用次数最少"为标准写入硬盘中存储。显示图片时也优先在内存中检查是否有缓存，如果没有再到硬盘中查找缓存。在退出页面时，react-native-fast-image并不急于清空内存中缓存的图片数据，而是在后面需要缓存新的图片时继续按照"最近一段时间内使用次数最少"原则进行逐步替换,也就是著名的LRU(Least Recently Used,最近最少使用)算法,这也是目前业界普遍使用的策略方案。

在移动设备硬件性能过剩的今天,react-native-fast-image使用的缓存策略可以更充分地利用硬件资源,相比Image可以减少闪烁(硬盘读写+解码的耗时),带来更好的用户体验。

5.2 音视频文件的操作方式

网易云音乐、豆瓣FM、抖音、快手、全民K歌……这些当前较为流行的娱乐社交平台都采用了音频或视频作为内容的主要承载方式。相比图文方式,音视频文件可以在短时间内承载更大的信息量,更容易获得用户的关注。但由于音视频文件体积较大,开发者在开发这类软件的需求时需要考虑文件的缓存、流媒体的压缩和解码、音视频文件操作等,其中最为基础的就是音视频文件的操作。

5.2.1 音频处理

React Native 官方并没有提供音频的处理能力,社区的开发者们先后推出了react-native-audio(仅录制)、react-native-sound(仅播放)等开源项目为iOS和Android的音频操作提供支持。react-native-audio-toolkit则吸收了这些开源项目的优点,集成了音频的录制和播放功能,并且为开发者提供了更为友好的调用方式,因此逐渐成为使用最为广泛的音频操作框架之一。

在了解react-native-audio-toolkit前,先简单介绍一下在原生端中处理音频的功能和方法有哪些。

1. iOS

iOS中的音视频功能由AVFoundation提供,该类为Apple官方类,但是默认不会引入,需要在使用的文件中手动引入。iOS主要提供了以下几个音频操作类。

AVAsset:一个抽象类,配置播放所需基本信息。

AVPlayer:实际音视频播放工具。

AVPlayerItem:AVPlayer媒体文件的载体,可以进行seek(快进/快退),并且提供播放状态的监听。

AVAudioRecorder:录音相关类。

常见的功能和方法有以下几种。

playerWithUrl:AVPlayer类方法,直接创建AVPlayer并且进行播放。

pause:暂停播放音频。

seekToTime:从指定时长位置开始播放。

如图5.6所示,react-native-audio-toolkit也主要是对这几个关键类进行了封装,结构如下。

其中直接暴露给JavaScript调用的有两个模块,即AudioPlayer和AudioRecorder,前者负责播放,后者负责录制。

AudioRecorder:简单封装维护了AVAudioRecorder供JavaScript端调用。

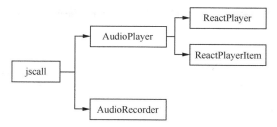

图5.6 react-native-audio-toolkit iOS 相关类

AudioPlayer：维护了ReactPlayer和ReactPlayerItem这两个类，并且它们分别继承于AVPlayer和AVPlayerItem。

整体来看，react-native-audio-toolkit并未基于原生做过多的改造，只是提供了基本能力的封装调用。

2. Android

Android中的音频操作主要使用MediaPlayer类完成，它可以接收媒体流作为数据源，并提供了一系列常用的媒体控制方法以及各个状态情况下的回调。在使用MediaPlayer时，首先要申请读写外部存储的权限。MediaPlayer提供的常用方法如下。

setDataSource：指定音频文件位置。

prepare/prepareAsync：播放之前完成准备工作。

start：开始或继续播放音频。

pause：暂停播放音频。

reset：将MediaPlayer对象重置到刚刚创建的状态。

seekTo：从指定位置开始播放。

stop：停止播放。调用后，MediaPlayer对象将无法再播放。

release：释放与MediaPlayer对象相关的资源。

isPlaying：判断是否正在播放。

getDuration：获取音频文件的时长。

录制音频可以使用MediaRecorder类，需要申请录制音频和读写外部存储的权限。MediaRecorder类提供的调用方法与前面使用的MediaPlayer基本一致，额外多出了几个录制时的常用方法，如以下几种方法。

resume：暂停录制后的继续录制。

setAudioSource：设置录制音频的来源，可以设置为主麦克风、通话的上下行音频、音频调试器的输出等。

setOutputFormat：设置输出的文件格式。

setAudioEncoder：设置音频编码方式。

setOutputFile：设置输出文件的路径。

react-native-audio-toolkit 基于 MediaPlayer 和 MediaRecorder 类完成了音频播放和录制功能的封装，具体的相关类如图 5.7 所示。

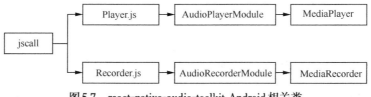

图 5.7　react-native-audio-toolkit Android 相关类

AudioPlayerModule：封装 MediaPlayer 的方法供 JavaScript 端调用。

AudioRecorderModule：封装 MediaRecorder 的方法供 JavaScript 端调用。

3. react-native-audio-toolkit

react-native-audio-toolkit 将主要的功能拆分为 Recorder（录音器）和 Player（播放器）两个模块，相比于原生音频操作的调用方法，react-native-audio-toolkit 提供的方法更为友好，比如支持本地文件名自动寻址，将播放/暂停与录制/停止分别合并为一个方法，多个音频文件在内存中的缓存处理等。如果想在项目中接入 react-native-audio-toolkit，那么就需要在 React Native 项目下的终端执行以下命令：

```
npm install --save @react-native-community/audio-toolkit
react-native link @react-native-community/audio-toolkit
```

由于音频插件所需用户权限的特殊性，引入代码后，开发者需要在各自平台上添加所需权限。

在 iOS 工程的 'Info.plist' 中添加：

```
<string>This app requires access to your microphone</string>
```

在 Android 工程的 'AndroidManifest.xml' 中添加：

```
<manifest ...>
    ......
    <!-- 读取外部存储公共目录的音频文件时需要添加 -->
    <uses-permission android:name="android.permission.READ_EXTERNAL_STORAGE" />
    <!-- 播放网络音频资源时需要添加 -->
    <uses-permission android:name="android.permission.INTERNET" />
    <uses-permission android:name="android.permission.ACCESS_NETWORK_STATE" />
    <!-- 录音时需要添加 -->
    <uses-permission android:name="android.permission.RECORD_AUDIO" />
    <!-- 将录音保存到外部存储公共目录中时需要添加 -->
    <uses-permission android:name="android.permission.WRITE_EXTERNAL_STORAGE" />
```

```
        ……
</manifest>
```

需要注意的是，如果Android项目已经使用了AndroidX的依赖库，那么此时可能还是无法顺利进行编译，因为目前react-native-audio-toolkit插件尚未适配AndroidX（编写本文时，react-native-audio-toolkit的版本为2.0.2）。如果此时你遇到这个问题，可以先查看一下react-native-audio-toolkit是否发布了适配AndroidX的版本。如果仍未适配，可以先在Android Studio右侧的Gradle中选择App并右击，选择Refresh Gradle project；然后在左侧同步出来的'react-native-community_audio-toolkit'项目中找到'AudioPlayerModule'和'AudioRecorderModule'，删除文件头部import依赖中的'import android.support.annotation.Nullable;'即可。具体位置如图5.8所示。

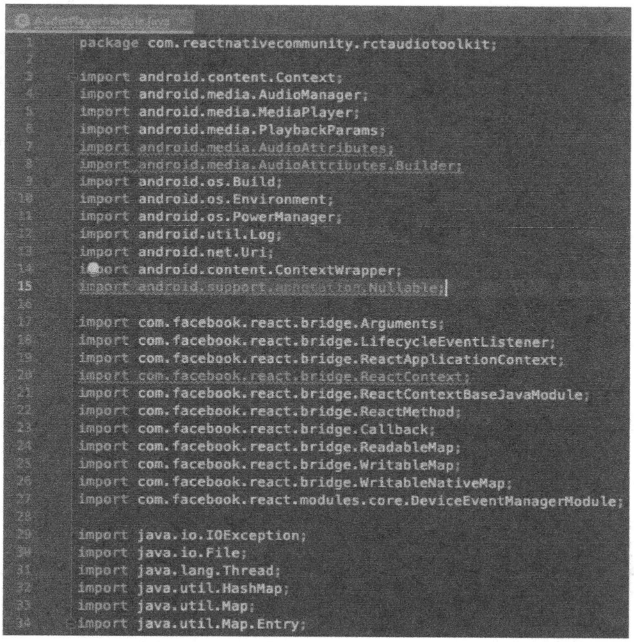

图5.8　AudioPlayerModule删除依赖位置

react-native-audio-toolkit提供的方法虽然在调用方式上很友好，但开发者需要注意的是这些方法的调用顺序与录音器和播放器的状态流转具有关联关系。图5.9中展示了播放器在不同调用方法执行时状态流转的情况。

播放器在初始化时，需要在构造方法中传入一个音频的路径。这个路径可以为本地文件名、本地文件绝对路径或网络音频URL。当播放器被实例化后，默认的状态为'IDLE'（空闲）状态；当调用'prepare()'方法时，状态变为'PREPARING'（准备中）。此时，播放器会加载音频文件的媒体流到内存，加载完毕后，状态变化为'PREPARED'（已准备）；此时可以执行'play()'方法开始播放音频，同时状态变为'PLAYING'（播放中）；如果在播放过程中调用'pause()'方法，则会进入'PAUSE'（已暂停）状态，想要继续播放，还需要调用一次'play()'方法；若在播放过程中调用'seek()'方法，则会进入'SEEKING'（快进/快退中）状态，执行完毕时会自动恢复到播放中的状态；音频在播放中、已暂停或快进、快退状态下执行'stop()'方法会让状态回到'PREPARED'（已准备）状态，音频正常播放完毕也会进入这个状态；另外，无论在什么运行状态下，调用'destroy()'方法都会让播放器进入'DESTROYED'（已销毁）状态；最终，无论在什么状态的流转过程中，只要出现方法调用时机不合理或调用过程失败的情况，都会进入'ERROR'（错误）状态。

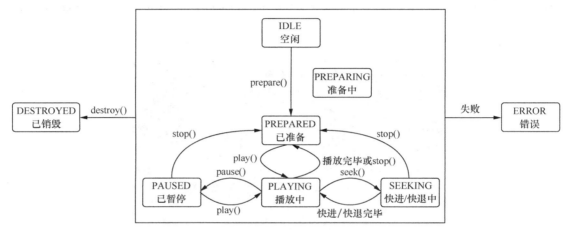

图5.9　播放器状态流转

录音器的调用方法和状态流转过程与播放器的相似，但由于减少了几个状态和方法，因此也存在一些差异。图5.10展示了录音器在不同调用方法执行时的状态流转。

录音器在构造方法中也需要传入一个音频文件的路径，这个路径只能是本地文件名或网络音频文件URL。当录音器被实例化后，默认的状态为'IDLE'（空闲）状态；调用'prepare()'方法的状态流转和播放器流程一致，但是在已准备状态下调用'record()'方法时，会进入'RECORDING'（录音中）状态；此时如果调用'pause()'方法则会使录音器进入'PAUSED'（已暂停）状态，并且录音器并没有提供方法继续录音；不论录音中还是已暂停状态都可以通过调用'stop()'方法，完成录制音频的保存并进入'DESTROYED'（已销毁）状态；无论在什么运行中的状态都可以通过调用'destroy()'方法使录音器不保存音频直接进入'DESTROYED'（已销毁）状态；无论在什么状态的流转过程中，只要出现方法调用时机不合理或调用过程失败的情况，都会进入'ERROR'（错误）状态。

上述的所有方法，都可以传入一个函数作为方法执行完毕时的回调，让开发者可以在每个方法执行完时更灵活地处理业务逻辑。梳理清楚了这些方法调用与播放器/录音器状态的流转的关系，下面就可以利用这些方法编写一个简易的录音/播放器，重点代码如下：

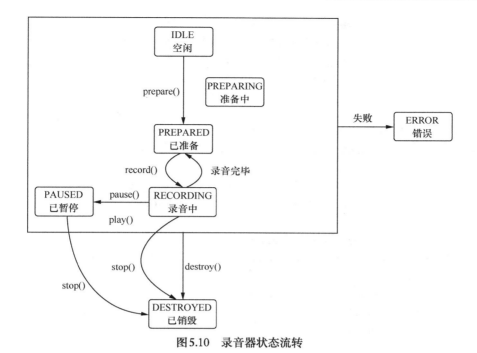

图5.10 录音器状态流转

```
import React, {Component} from 'react';
import {StyleSheet, View, Button, Text, Slider, Switch, PermissionsAndroid, Platform}
from 'react-native';
import { Player, Recorder } from '@react-native-community/audio-toolkit';

const MUSIC_URL = ''; // 网络音频资源
const MUSIC_FILE_NAME = 'RNAudioTest.mp4'; // 本地音频资源

export default class App extends Component {
  player: Player | null; // 播放器
  recorder: Recorder | null; // 录音器
  lastSeek: number; // 记录用户上次拖动进度条的时间,用于避免用户拖曳进度条与播放时更新进度
条同时更改进度条位置发生冲突
  _progressInterval = null; // 更新进度条的逻辑

  constructor(props) {
    super(props);
    this.state = {
      duration: '-- : --', // 默认的音频持续时长
      progress: -1, // 进度条进度初始值
```

```js
      mediaStatus: '加载中', // 播放器状态初始值
      useUrl: false, // 控制开关状态,是否切换使用网络音频资源
      audioSource: MUSIC_FILE_NAME, // 音频资源,默认使用本地音频资源
    };
  }
  ......
  // 请求录音权限(Android)
  async _requestRecordAudioPermission() {
    try {
      const granted = await PermissionsAndroid.request(
        PermissionsAndroid.PERMISSIONS.RECORD_AUDIO,
        {
          title: 'Microphone Permission',
          message: 'ExampleApp needs access to your microphone to test react-native-audio-toolkit.',
          buttonNeutral: 'Ask Me Later',
          buttonNegative: 'Cancel',
          buttonPositive: 'OK',
        },
      );
      return granted === PermissionsAndroid.RESULTS.GRANTED;
    } catch (err) {
      console.error(err);
      return false;
    }
  }
  ......
  // 开始录音
  startRecord = () => {
    if (this.player) { // 录音与播放不能同时启用(因为本例中操作的是同一个本地文件)
      this.player.destroy();
    }
    // 请求录音权限
    let recordAudioRequest;
    if (Platform.OS == 'android') {
      recordAudioRequest = this._requestRecordAudioPermission();
    } else {
      recordAudioRequest = new Promise(function (resolve, reject) { resolve(true); });
    }
```

```
  recordAudioRequest.then((hasPermission) => {
    if (!hasPermission) {
      console.warn('获取录音权限失败');
      return;
    }
    // 请求权限成功，开始录音
    this.recorder.record((err) => {
      if (err) {
        console.warn(err.message);
        return;
      }
      this.setState({mediaStatus: '录音中'});
      console.log(this.player.state);
    });
  });
};
......
}
```

实际效果如图5.11所示。

图5.11　react-native-audio-toolkit实际效果

从上面的示例中可以看出，react-native-audio-toolkit对iOS和Android文件系统的差异做了一定的处理，在设置本地音频文件资源时，可以只设置一个文件名，而不用关心具体的目录。在录音时，Android会将录音文件保存至'内部存储/Package目录/files'目录下，iOS则存放在Document目录下。在播放时，由于JavaScript端只提供了一个文件名，Android会依次去'内部存储/Package目录/files'目录、外部存储根目录寻找音频文件；若未找到，则查看传入的路径是否为绝对路径；最后还会到App打包的资源文件中寻找音频文件。iOS则默认会从Document目录下寻找，如果文件不存在则会继续从App资源包（mainBundle）中查找。这使开发者不用关心两个文件系统的差异性，但也损失了一定的灵活性。

示例中使用了播放器的'play()/pause()'方法和录音器的'record()/stop()'方法，完成了音频的播放/暂停和录制/停止。考虑到一些场景下，前端的设计要求播放器的播放按钮在点击后变成暂停按钮，或者录音器的录音按钮在点击后变成停止按钮，react-native-audio-toolkit为了防止开发者在前面叙述的状态控制与方法调用中出现更多的问题，提供了'playPause()'和'toggleRecord()'方法来控制播放/暂停和录制/停止的逻辑。这样，开发者在调用这些方法时传入一个回调更改按钮的样式和文字，就可以快速实现这样的前端需求。

本例在播放器和录音器的构造方法中，传入了音频资源的路径，得到了与该音频资源关联的播放器/录音器实例。如果需要加载多个音频资源准备播放时，可以构造出多个播放器/录音器实例，在原生端，react-native-audio-toolkit按照这些实例的构造顺序维护了一个播放器/录音器池。这样就可以同时在内存中维护多个音频资源的播放器/录音器，满足更为复杂的需求。

示例中还用到了获取播放器/录音器的基本信息（duration、currentTime等），没有涵盖到的信息还有loop（循环模式）、volume（音量）、speed（倍速）等，开发者可以在实现特定需求时，获取这些信息进行处理。同时，示例也表现出了react-native-audio-toolkit的一些缺陷，比如录音只有暂停录制的方法没有继续录制的方法，维护多音频播放器/录音器实例时都保存在内存中可能占用内存过多等，这些问题还需要继续完善。

5.2.2 视频处理

react-native-video是目前社区较为流行的视频播放组件，它支持播放网络视频和本地视频资源，提供了视频播放基本的播放控制、倍速、音量调节、画面缩放模式等功能，对于iOS还有视频离线缓存等功能。

同音频一样，下面先简单介绍一下原生端中处理视频的相关类及功能。

1. iOS

iOS中的视频处理功能与音频类似，主要依赖于上文提到的AVAsset、AVPlayer和AVPlayerItem 3个大类。在react-native-video中，视频基本的播放/暂停、快进、视图大小切换，以及旋转屏适配，均由原生API和原生视图提供，相关类关系如图5.12所示。

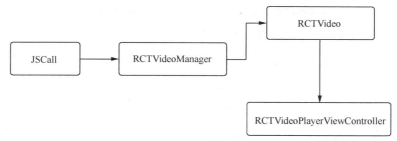

图 5.12　react-native-video iOS 相关类

RCTVideoManager：原生端 JavaScript 的导出类，提供了视图对象 RCTVideo 给 JavaScript 使用，并提供 start、end 等音频的交互功能。

RCTVideo：负责通知 ViewManager 各种事件（例如 onStart 和 onEnd）集成与 UIView，内部管理 AVPlayer 和 AVPlayerItem。

RCTVideoPlayerViewController：继承于 AVPlayerViewController；AVPlayerViewController 是原生提供的最简单的视频 ViewController，提供了最基本的视频操作及 UI 交互。RCTVideo 将其管理的 AVPlayer 直接赋给 RCTVideoPlayerViewController 执行播放操作。

由于继承于 React Native 的 ViewManager（RCTVideoManager）的类必须提供一个 UIView 对象给 JavaScript，但这里实际用到的播放对象是一个 ViewController（RCTVideoPlayerViewController），因此中间又提供了一个桥接类 RCTVideo 来处理这个问题。该类并不会使用 UIView 的任何方法，而是直接把 manager 中 JavaScript 配置的各种属性和操作设置给自身持有的 AVPlayer 对象，然后通过赋值的方式重置 RCTVideoPlayerViewController 中的 AVPlayer 对象，并且调起 ViewController 进行播放。

2. Android

在 Android 中也可以通过上面提到的 MediaPlayer 和 MediaRecorder 进行视频的播放和录制，同时也有不少其他相关的类来帮助我们更好地操作视频。

通常情况下可以采用 Camera+SurfaceView+MediaRecorder 的方案开发视频录制的功能，其中 Camera 是 Android 提供的操作相机硬件的类，SurfaceView 可以用来将相机拍摄到的画面展示到屏幕中，MediaRecorder 则主要完成对视频文件的录制和存储。

MediaRecorder 进行相机录制时虽然较为简便，与录制音频的调用方法基本一致，但在录像开始和结束时会发出提示音，并且不能去除。这是谷歌为了防止开发者开发用于偷拍的 App 采取的反制措施。如果你需要更高阶的需求，可以考虑使用开源框架 FFMPEG，大部分的 App 都是基于 FFMPEG 实现的视频录制和视频处理。

另外，在视频播放方案上 Android 还在 MediaPlayer 的基础上做了进一步封装，提供了调用更为简单的 VideoView 供开发者使用。同时，开发者直接使用 Android 提供的播放控制组件 MediaController，无须开发播放、暂停按钮和进度条等 UI，就能直接快速实现视频播放的功能。

不过react-native-video并没有采用原生系统封装好的VideoView和MediaController，而是自己基于MediaPlayer和TextureView封装了一个视频组件，具体的结构如图5.13所示。

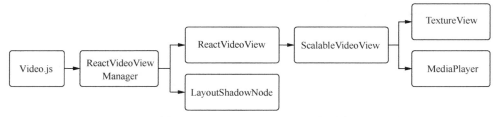

图5.13　react-native-video Android 相关类

ReactVideoViewManager：用于将ReactVideoView和LayoutShadowNode绑定，并且接收JavaScript端使用组件时设置的一系列参数，其中布局相关的参数设置给LayoutShadowNode，UI交互相关的参数设置给ReactVideoView。

LayoutShadowNode：接收布局相关参数，基于Yoga框架实现了Flex布局。

ReactVideoView：基于ScalableVideoView封装的可供React Native使用的视频组件，主要负责向React Native回调UI事件，解析播放源地址等。

ScalableVideoView：基于TextureView和MediaPlayer封装的视频播放组件。

TextureView：用于显示视频画面，支持缓冲、缩放、旋转，相比SurfaceView在UI层级上具有更好的管理，但更加消耗内存。

MediaPlayer：用于加载媒体资源，控制播放、暂停、跳转等媒体操作。

3. react-native-video

react-native-video的安装方式与其他插件类似。

```
npm install --save react-native-video
react-native link react-native-video
```

基本用法如下。

```
// 根据 videoSource 的状态切换本地资源和网络资源
get displaySource() {
    const { videoSource } = this.state;
    // 本地资源需要用 require 引入，网络资源需要用对象的 uri 字段引入
    return videoSource === 'localVideo' ?
        require('./broadchurch.mp4') :
        { uri: 'https://sjycdn.miaopai.com/stream/HNkFfNMuhjRzDd-q6j9qycf540aKqInVMu0YhQ___8.mp4?ssig=5cc4461ddaf237f6a4b9cf5ec677ad36&time_stamp=1578383924657',
type: 'mpd' };
```

```
        }

        // 播放完毕后将播放状态变为暂停并将进度条归零
        onEnd = () => {
            this.setState({ paused: true })
            this.video.save(); // 视频离线缓存，仅对iOS有效
            this.video.seek(0); // 进度跳转
        };

        render() {
            return (
                <Video
                    ref={(ref: Video) => { this.video = ref }}
                    source={this.displaySource} // 设置视频资源
                    style={styles.fullScreen}
                    rate={this.state.rate} // 播放倍速，浮点数，1.0 为正常倍速
                    paused={this.state.paused} // 播放状态，布尔值，true 为暂停，false 为播放
                    volume={this.state.volume} // 音量控制，浮点数，1.0 为视频资源默认音量
                    muted={this.state.muted} // 静音状态，布尔值，true 为静音
                    resizeMode={this.state.resizeMode} // 画面缩放模式，'contain' 居中，'cover' 平铺，'stretch' 拉伸
                    onLoad={this.onLoad} // 加载资源结果回调
                    onProgress={this.onProgress} // 播放进度回调
                    onEnd={this.onEnd} // 播放完毕回调
                    onAudioBecomingNoisy={this.onAudioBecomingNoisy} // 从耳机到扬声器切换时的回调
                    // 播放通道被其他软件的播放源抢占时的回调
                    onAudioFocusChanged={this.onAudioFocusChanged}
                    repeat={false}
                />
            )
        }
```

实际效果如图5.14所示。

图5.14　react-native-video效果

5.3 本地文件系统

iOS和Android的本地文件系统是平台差异较大的一个方面：iOS采用封闭的沙盒机制，App没有权限读写不在其目录下的任何文件；而Android也出于对安全和隐私的考虑，经过几个系统版本的迭代，逐渐向iOS的方案靠拢，App可以不用权限直接访问其在外部存储中的Package目录，而访问外部存储的其他公共目录则需要读写外部存储的权限。本节将介绍iOS和Android的本地文件系统，以及目前React Native对其提供的方案。

5.3.1　iOS本地文件系统

iOS项目中通常会具备以下几个目录。

Documents：一般用于存放重要的文件信息，如sqlite等。

Library：包含Caches和Preferences两个子目录，Caches用于存放临时缓存（包括图片缓存），Preferences用于存放用户的偏好设置。

tmp：用于存放临时信息，当系统认为需要释放手机空间时，就会清除App在tmp目录下的文件。

具体的结构如下。

```
App 目录
    ├── Documents
    ├── Library
    │    ├── Caches
    │    └── Preferences
    └── tmp
```

5.3.2 Android本地文件系统

目前Android的本地文件系统分为Internal Storage（内部存储）和External Storage（外部存储），其中内部存储的文件目录主要是系统负责维护，供开发者使用的主要是外部存储。

```
内部存储
    └── data
         └── data
              └── com.xxx.xxx
                   ├── shared_prefs
                   ├── databases
                   ├── files
                   └── cache

外部存储
    ├── Music (Podcasts/Ringtones/Alarms/Notifications)
    ├── Pictures
    ├── Movies
    ├── Download
    ├── DCIM
    │    └── Screenshots
    ├── Documents
    ├── Audiobooks
    └── Android
         └── data
              └── com.xxx.xxx
                   ├── cache
                   └── files
```

内部存储的Package目录通常位于data/data/下，目录名称为App的包名（通常为倒序的域名），并且具有二级目录shared_prefs、databases、files、cache，分别用来存储SharedPreference信息、SQLite数据库信息、文件信息和缓存信息。

外部存储则包括Music（通常为Podcasts/Ringtones/Alarms/Notifications的合集）、Pictures、Movies、Download、DCIM、Documents、Audiobooks这几个公有目录，通常用来存储音乐、图片、电影、下载文件、照片视频或截屏、文档、有声读物等文件，其中Screenshots通常作为二级目录被放置在Pictures或DCIM目录下。

访问公有目录需要在Manifest文件中声明读写外部存储的权限，并在运行时动态请求权限。访问外部存储中的Package目录则不需要权限，谷歌鼓励开发者将App相关的数据和缓存都存放在这里，这样卸载App时可以将这些数据全部清除。Package目录位于外部存储的 Android/data/ 路径下，目录名称为App的包名（通常为倒序的域名），Package目录中一般具有二级目录cache和files，用于存放缓存和文件。

5.3.3　react-native-fs

同音视频一样，React Native官方也没有提供能够操作本地文件的功能模块，而react-native-fs则是目前社区较为流行的操作本地文件系统的自定义插件。它提供了一些大部分常规的文件及文件夹操作，且对于不同平台的差异，也提供了一些各平台独有的API。

react-native-fs主要包括下面这些功能。
- 创建文件/文件夹；
- 删除文件/文件夹；
- 更新文件；
- 读取文件；
- 移动文件/文件夹；
- 复制文件/文件夹；
- 获取文件简介；
- 设置文件创建/修改时间；
- 获取文件系统信息（存储使用情况）；
- 判断文件是否存在；
- 上传/下载文件。

react-native-fs中的方法基本支持Promise方式，以下是文件写入和读取的调用示例代码。

```
import React, {Component} from 'react';
import {View, Button, Text, ScrollView} from 'react-native';
import RNFS from 'react-native-fs';

export default class App extends Component {
```

```
state = {
  dirFileList: [] // 目录中的文件列表
};

// RNFS.DocumentDirectoryPath 是 iOS 和 Android 两个平台均有的目录地址
// 使用 UTF-8 编码
dir = RNFS.DocumentDirectoryPath;
path = this.dir + '/test.txt';
encode = 'utf8';
subDir = '/subDir';

/**
 * 创建文本文件
 */
createTxtFile = () => {
  // 要写入的内容
  const content = '这是 txt 文件的内容';
  RNFS.writeFile(this.path, content, this.encode)
    .then(success => {
      console.log('文件已写入！');
    })
    .catch(err => {
      console.error(err);
    });
};


/**
 * 读取文本文件
 */
readTxtFile = () => {
  // 判断文件是否存在
  RNFS.exists(this.path)
    .then(isExists => {
      if (!isExists) {
        console.warn('文件不存在，请先创建！')
        return;
      }
      // 读取文件内容
      RNFS.readFile(this.path, this.encode)
        .then(content => {
```

```javascript
                console.warn('读取到了文档内容: ', content);
            })
            .catch(err => {
                // 也可以不用RNFS.exists检查文件是否存在
                // 如果文件不存在,可以在此处接收错误信息并进行逻辑处理
                this.warnErr(err);
            });
        })
        .catch(err => {
            console.error(err);
        });
};

/**
 * 获取文件简介
 */
getFileInfo = () => {
    RNFS.stat(this.path)
        .then(result => {
            const fileInfo =
                '路径: ' + result.path + '\n' +
                '创建时间: ' + result.ctime + '\n' +
                '修改时间: ' + result.mtime + '\n' +
                '占用空间: ' + result.size + ' B\n' +
                '读写类型: ' + result.mode + '\n' +
                '原始路径: ' + result.originalFilepath + '\n' +
                '是否为文件: ' + result.isFile() + '\n' +
                '是否为文件夹: ' + result.isDirectory();
            console.warn(fileInfo);
        })
        .catch(err => {
            this.warnErr(err);
        });
};
}
```

需要注意的是,react-native-fs提供的目录常量在不同平台上所指的具体路径不同,例如上边使用的操作目录为'RNFS.DocumentDirectoryPath',在iOS中这个常量的地址为沙盒中的Documents目录,而在Android中这个常量的地址为内部存储中Package目录下的files目录。除此之外,react-native-fs还提供了许多常量表示两个平台的常用目录,表5.1可以快速地帮助你弄清这些常量在不同平台究竟指的

是哪个目录。

表5.1

react-native-fs 目录常量名	对应 iOS 目录	对应 Android 目录
MainBundlePath	AppName.app 程序包目录	iOS 特有
CachesDirectoryPath	沙盒 /Library/Caches	内部存储 /Package 目录 /cache
ExternalCachesDirectoryPath	Android 特有	外部存储 /Package 目录 /cache
DocumentDirectoryPath	沙盒 /Documents	内部存储 /Package 目录 /files
ExternalDirectoryPath	Android 特有	外部存储 /Package 目录 /files
ExternalStorageDirectoryPath	Android 特有	外部存储根目录
TemporaryDirectoryPath	沙盒 /tmp	内部存储 /Package 目录 /cache（与 CachesDirectoryPath 一致）
LibraryDirectoryPath	沙盒 /Library	iOS 特有
PicturesDirectoryPath	Android 特有	外部存储 /Pictures（或外部存储 /DCIM/Pictures）

5.4 本地存储

Web 开发和移动端开发存在一个很重要的区别就是本地数据持久化的能力。Web 系统很大程度上会依赖网络的状况，一旦网络出现异常，服务基本也就丧失了功能；而移动端作为端设备，本身具备一定的本地存储能力，开发得当的话在弱网环境下也可以进行良好的操作，并等待网络恢复后再将操作产生的影响同步到服务端。虽然随着网络越来越便捷，大部分 App 处于没有网络就无法使用的情况，但对于一些工具类 App，例如文档、记账等类型，本地存储或者离线存储仍是一个非常重要的场景。本节会介绍一下 iOS 和 Android 下的本地存储能力，以及 React Native 目前可以使用其中的哪些功能。

5.4.1 iOS 本地存储方式

iOS 主要支持以下几种存储方式。

plist：本质是把数据以 xml 的形式存储在本地文件中，需要自己创建文件并且写入本地沙盒。

NSUserDefault：一个本地系统预先创建好的文件，读写方式非常简单方便，但需要注意的是如果不调用 synchronize 方法，系统会根据 I/O 情况自动分配保存到文件中的时间。

NSKeyedArchiver：只支持简单对象的直接存储（如原生提供的数据类型），当在业务中使用自定义 model（模型）时，需要给该 model 自定义编码（NSCoding）进行归档和反归档才能正常存储，如果需要归档的类是某个自定义类的子类时，在归档和解档之前需要先实现父类的归档和解档方法。

CoreData 和 SQL：标准的数据库存储。

5.4.2 Android本地存储方式

Android提供的数据持久化方式分为SharedPreferences、SQLite、文件存储和ContentProvider几种，它们的能力各不相同，可以应对不同的业务场景。下面将简述这几种数据持久化方式的特点和使用方式。由于Android自身是开源的关系，我们能更清晰地看到它的实现方式。

1. SharedPreferences

SharedPreferences是一种轻量级的本地存储方式，采用Key-Value（键值对）的方式进行存储，一般用于存储一些简单且常用的参数配置信息。它的本质是将要存储的数据及其类型保存成XML文件，并存放在App的Package中的shared_prefs目录下。

下面的示例展示了SharedPreferences的基本调用方式，存储了用户的基本信息。

```java
// SharedPreferences 的名称
String spName = "SPTest";
// 从上下文对象 context 中获取 SharedPreferences 对象
SharedPreferences sp = context.getSharedPreferences(spName, Context.MODE_PRIVATE);
// 写入
sp.edit()
    .putString("userName", "小明")
    .putInt("userAge", 18)
    .putBoolean("isVip", false)
    .apply();
// 读取，所有 get 方法的第一个参数为 Key 值，第二个参数为未找到 Value 时返回的默认值
String userName = sp.getString("userName", "默认名称");
int userAge = sp.getInt("userAge", -1);
boolean isVip = sp.getBoolean("isVip", false);
```

在执行完上面的示例代码后我们可以在App的'内部存储/Package目录/shared_prefs'路径下找到生成的XML文件，如图5.15所示。

图5.15　SharedPreferences存储路径

结合示例，我们可以总结出SharedPreferences具有如下特点。

（1）轻量级存储，适用于小规模的数据类型。

（2）可以在存储时区分数据类型。

（3）将数据以XML文件的形式存储在App Package目录下，卸载App后数据随之清除。

（4）一个SharedPreferences对象对应一个XML文件，读取时会将整个XML文件的所有Key-Value都读到内存中并转化为一个Map对象，即使用户只需要其中一个Key-Value数据。

2. SQLite

SQLite 是一个轻量级的关系型数据库,支持常见的语言和操作平台,运算速度很快,占用资源极少,它没有用户密码和管理员等数据库管理的概念,提供了最纯粹的关系数据存取的功能,兼容标准 SQL 语法。上文的示例用 SharedPreferences 存储了一个用户的基本信息。可以看出用户名称、年龄、是否为会员这样的数据之间具有一定的关系,它们描述的是同一个用户的不同属性。因此根据需求,我们也可以用 SQLite 将这一条用户的基本信息整体存入用户数据表中,和其他用户信息进行区分,便于后续的用户信息查询操作。

下面的示例展示了使用 SQLite 数据库存取一条用户基本信息的常见调用方式。

```java
public class MyDBHelper extends SQLiteOpenHelper {
    public MyDBHelper(@Nullable Context context, @Nullable String name, @Nullable SQLiteDatabase.CursorFactory factory, int version) {
        super(context, name, factory, version);
    }

    @Override
    public void onCreate(SQLiteDatabase db) {
        // 这里进行建表操作
        db.execSQL("create table User (" +
                "id integer primary key autoincrement," +
                "username text," +
                "userage integer," +
                "isvip boolean)"
        );
    }

    @Override
    public void onUpgrade(SQLiteDatabase db, int oldVersion, int newVersion) { }
}

// 新建数据库工具类,创建名称为 SQLitTest 的数据库
MyDBHelper mDBHelper = new MyDBHelper(context, "SQLiteTest", null, 1);
// 插入数据
SQLiteDatabase writableDatabase = mDBHelper.getWritableDatabase();
ContentValues values = new ContentValues();
values.put("username", "小明");
values.put("userage", 18);
values.put("isvip", 0);
writableDatabase.insert("User", null, values);
```

```
    // 查询数据
    Cursor cursor = writableDatabase.query("User", null, null, null, null, null, null);
    if (cursor.moveToFirst()) {
        do {
            String userName = cursor.getString(cursor.getColumnIndex("username"));
            int userAge = cursor.getInt(cursor.getColumnIndex("userage"));
            boolean isVip = cursor.getInt(cursor.getColumnIndex("isvip")) == 1;
        } while (cursor.moveToNext());
    }
    cursor.close();
```

运行完以上的示例代码后，可以在'内部存储/package目录/databases'下找到刚刚创建出来的SQLiteTest数据库文件（见图5.16），这些是SQLite3类型的二进制文件，需要使用adb shell指令，或导出该文件后使用第三方软件查看。

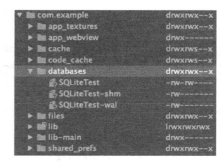

图5.16　SQLite存储路径

综上所述，SQLite具有以下特点。

（1）调用方式严谨复杂，支持SQL语法。

（2）具有版本控制。

（3）可以满足复杂的查询操作。

（4）相比于其他数据库，占用内存较小，查询速度较快。

3. 文件存储

前面的两种本地存储方式实质上都是经过封装的文件存储，SharedPreferences是将数据以XML文件的格式进行保存，SQLite是将数据以SQLite3格式的database文件进行存储。然而一些情况下需要将特定格式的数据进行存储，例如缓存广告图片、缓存提示音文件或者将数据按加密算法编码后存储等，此时就需要直接对文件写入和读取。

Android读写文件的操作基本和Java常见的文件操作一致，唯一需要注意的是读写外部存储的公共目录需要权限，这部分数据不会随着App的卸载而删除；内部/外部存储的Package目录下进行文件的读写不需要申请权限，但App卸载后这部分数据会随之清除。

下面的示例展示了Android在内部存储的Package目录中进行文件读写的代码。

```
    String PACKAGE_FILE_DIR = context.getFilesDir().getAbsolutePath(); // 获取内部存储/Package目录/files目录路径
    String FILE_NAME = "file_test.text"; // 要存储/读取的文件名
    // 构造要存储的文件内容
    String content = "userName: 小明\n" +
        "userAge: 18\n" +
        "isVip: false";
```

```java
// 使用输出流写入文件
File file = new File(PACKAGE_FILE_DIR, FILE_NAME);
FileOutputStream outputStream = null;
try {
    outputStream = new FileOutputStream(file);
    outputStream.write(content.getBytes()); // 将文件的字节码写入
} catch (IOException e) {
    e.printStackTrace();
} finally {
    if (outputStream != null) {
        try {
            outputStream.close();
        } catch (IOException e) {
            e.printStackTrace();
        }
    }
}
// 使用输出流读取文件
content = null;
FileInputStream inputStream = null;
try {
    inputStream = new FileInputStream(file);
    byte[] buf = new byte[1024]; // 1KB的读取缓存
    StringBuilder sb = new StringBuilder();
    while (inputStream.read(buf) != -1) { // 循环读取
        sb.append(new String(buf).trim()); // 拼接内容
    }
    content = sb.toString(); // 将拼接的内容进行赋值
} catch (IOException e) {
    e.printStackTrace();
} finally {
    if (inputStream != null) {
        try {
            inputStream.close();
        } catch (IOException e) {
            e.printStackTrace();
        }
    }
}
System.out.println(content == null ? "读取文件失败" : content); // 输出读取的文件内容
```

上述代码使用文件读写/存储了一个txt文件，并把用户信息字符串以字节码的形式存储，执行后，我们可以在要写入的目录下找到刚刚存储的txt文件，如图5.17所示。

图5.17　文件存储路径

文件的写入和读取是最原始和灵活的本地存储方案，基于文件存储可以封装出适用于各种各样业务场景的存储方案。但可以看出文件存储的逻辑较为复杂，需要用try-catch处理各种异常情况。而且如果没有分级缓存的读写策略，那么直接在硬盘中的读写I/O开销很耗费性能。Android要求开发者必须在子线程中完成文件的读写操作，防止过长的读写时间阻塞UI线程，影响用户体验。

4. ContentProvider

ContentProvider严格来讲是数据持久化的一种数据访问方式，而不是一种存储方式。作为Android的四大组件之一，ContentProvider主要负责App间的数据交互，保障系统的安全性和数据的隐私性。

Android希望App尽量只访问自己Package目录下的资源，在用户授权允许的情况下可以访问公共资源。那么如果某个App想要访问别的App存储的数据该怎么办？比如你的App想在通信录（又称通讯录）App中选取一个联系人并将这个联系人的信息回传，或者你希望本公司的其他App可以从你负责的App的用户列表中选取一个用户并将用户信息传递过去。

如果每个App都能够直接读写别的App下存储的数据，势必会带来灾难性的安全和隐私问题。考虑到App间的数据共享问题，Android在其他存储方式的基础上进一步封装，推出了ContentProvider这样的概念。

ContentProvider将需要共享的数据存储后，生成一个URI（统一资源标识符），这样其他App就可以通过这个URI完成对共享数据的访问，并且这样的访问是在ContentProvider提供的方法下进行的，更加提升了封装性和安全性。

5.4.3　React Native本地存储方式

介绍完iOS和Android原生的本地存储方式，相信你对移动端的能力也有了一定的了解，那React Native又提供了怎样的本地存储能力？最早官方内置了AsyncStorage模块，数据以Key-Value的形式存储。在底层实现上，AsyncStorage会根据数据量的规模进行判断，数据量较小时，在iOS使用序列化的字典存储，在Android使用Facebook自己的RocksDB存储；数据量较大时，在iOS会写入单独的文件，在Android使用SQLite存储。

后来Facebook官方宣布不再维护AsyncStorage，转而推荐使用社区的react-native-community/react-native-async-storage作为解决方案。目前，社区版本与AsyncStorage模块的区别在于Android移除了数据规模的判断，舍弃了RocksDB，直接使用SQLite进行存储，并且需要像引入自定义模块那样来安装react-native-async-storage，例如：

```
npm install @react-native-community/async-storage --save
react-native link @react-native-community/async-storage
```

AsyncStorage和react-native-async-storage的API基本保持一致，大部分的接口也均为异步接口，所以既支持参数传递回调函数，也支持返回Promise对象，或者可以使用async/await模式，例如：

```
import React, { Component } from 'react';
import { View, Button } from 'react-native';
import AsyncStorage from '@react-native-community/async-storage';

export default class App extends Component {

  // 示例：保存用户信息
  storeUser = async () => {
    const userString = JSON.stringify({
      userName: '小明',
      userAge: 18,
      isVip: false
    });

    const result = await AsyncStorage.setItem('user', userString);
    console.log('user saved:', result);

    AsyncStorage.setItem('user1', userString, error => {
      console.log('user1 saved: ', error);
    });

    AsyncStorage.setItem('user2', userString).then(error => {
      console.log('user2 saved', error)
    })
  };

  // 示例：读取用户信息
  getUser = async () => {
    const result = await AsyncStorage.getItem('user');
```

```
      console.log('user', result);

      AsyncStorage.getItem('user1', (error, result) => {
        console.log('user1', result)
      });

      AsyncStorage.getItem('user2').then(result => {
        console.log('user2', result)
      });
    };

    render() {
      return (
        <View style={{ flex: 1, justifyContent: 'center', alignItems: 'center' }}>
          <Button title={'保存用户信息'} onPress={this.storeUser} />
          <Button title={'读取用户信息'} onPress={this.getUser}/>
        </View>
      );
    }
  }
```

除了读写操作外,async-storage模块提供的常用接口还包括以下几种。

mergeItem:合并两个JSON类型字符串的值,类似于Object.assign(value1, value2)。

```
// 使用方法如下
await AsyncStorage.setItem('user', JSON.stringify(user))
await AsyncStorage.mergeItem('user', JSON.stringify(user))
```

removeItem:删除指定Key的值。

getAllKeys:获取所有设置的Key。

multiGet/multiSet/multiMerge/multiRemove: getItem/setItem/mergeItem/removeItem 的批量操作,唯一需要注意的是批量操作时采用二维数组标识Key-Value关系,而不是对象,例如:

```
await AsyncStorage.multiSet([['user', 'user_value'], ['user1', 'user1_value']])
```

5.4.4　React Native混合模式下的公共存储方案

目前,大部分React Native的自定义模块方案都会倾向于纯粹的Web/JavaScript开发者,直接提供接口来使用原生的能力。但在实际场景中,我们很可能会需要在原生项目中内嵌一个React Native项目,并且原生和JavaScript之间需要互传、共享一部分数据(例如用户信息)。在这种混合模式下，AsyncStorage存储的内容就没有办法直接被原生端使用，你需要阅读AsyncStorage模块的源码，找到它使用存储的具体本地位置，再编写对应的原生代码来进行获取。这个方式相对烦琐，毕竟AsyncStorage主要应对的场景都是React Native端的读写，并不考虑一端读、另一端写的情况。其实可以提供一个面向两端的存储方案，除了给React Native提供读写的API外，也可以给原生端提供读写的接口，划出一块公共的区域给原生和JavaScript共享其中的数据，避免因业务场景不同而提供不同的数据读写接口，大致结构如图5.18所示。

图5.18　混合存储模块设计

实现的代码比较简单，下面先采用最简单的存储方式做一个说明。

1. 原生端Storage类

首先在iOS和Android中实现一个专门用于存储的Storage类，下面的例子简单实现了字符串形式的Key-Value的读写。

```objc
// iOS
#import "MCRNStorageCore.h"

NSString *MCRNHandleKeys(NSString *key) {
    return [NSString stringWithFormat:@"MCRNKey-%@", key];
}

@interface MCRNStorageCore ()

@property(nonatomic, strong) NSUserDefaults *userDefaults;

@end

@implementation MCRNStorageCore

- (NSUserDefaults *)userDefaults {
    return [NSUserDefaults standardUserDefaults];
}

+ (instancetype)shared {
    static id obj = nil;
    static dispatch_once_t onceToken;
```

```objc
        dispatch_once(&onceToken, ^{
            obj = [[self alloc] init];
        });
        return obj;
}

- (void)setItem:(NSString *)item forKey:(NSString *)key {
    if (item && key.length) {
        [self.userDefaults setValue:item forKey:MCRNHandleKeys(key)];
        [self.userDefaults synchronize];
    }
}

- (nullable NSString *)getItemForKey:(NSString *)key {
    if (key.length) {
        return [self.userDefaults valueForKey:MCRNHandleKeys(key)];
    }
    return nil;
}

@end
```

```java
// Android
package com.meicai.react.storage;

import android.content.Context;
import android.content.SharedPreferences;

import com.facebook.react.bridge.ReadableArray;
import com.facebook.react.bridge.WritableArray;
import com.facebook.react.bridge.WritableNativeArray;

import java.util.ArrayList;
import java.util.List;

public class MCRNStorage {
    public static final String mTAG = "mc_storage";
    // 创建一个写入器
    private static SharedPreferences mPreferences;
    private static SharedPreferences.Editor mEditor;
```

```java
private static volatile MCRNStorage sInstance;

// 构造方法
public MCRNStorage(Context context) {
    mPreferences = context.getSharedPreferences(mTAG, Context.MODE_PRIVATE);
    mEditor = mPreferences.edit();
}


public static MCRNStorage getInstance(Context context) {
    if (sInstance == null) {
        synchronized (MCRNStorage.class) {
            if (sInstance == null) {
                sInstance = new MCRNStorage(context);
            }
        }

    }
    return sInstance;
}

// 存入数据
public void setItem(String key, String value) {
    mEditor.putString(key, value);
    mEditor.commit();
}

// 获取数据
public String getItem(String key) {
    return mPreferences.getString(key, "");
}

}
```

2. React Native自定义模块

上述Storage类可被再次封装成React Native的自定义模块,暴露API给JavaScript,大致的关键代码如下。

```
// iOS
#import "MCRNStorageAPI.h"
#import "MCRNStorageCore.h"
```

```objc
#import <React/RCTBridgeModule.h>

@interface MCRNStorageAPI ()<RCTBridgeModule>

@end

@implementation MCRNStorageAPI

RCT_EXPORT_MODULE(MCRNStorage)

- (dispatch_queue_t)methodQueue {
    return dispatch_queue_create("MCRNStorageAPIQueue", DISPATCH_QUEUE_SERIAL);
}


RCT_EXPORT_METHOD(setItem:(NSString *)itemKey andItem:(NSString *)item) {
    [[MCRNStorageCore shared] setItem:item forKey:itemKey];
}

RCT_EXPORT_METHOD(getItem:(NSString *)itemKey
                  resolver:(RCTPromiseResolveBlock)resolve
                  rejecter:(RCTPromiseRejectBlock)reject) {
    NSString *item = [[MCRNStorageCore shared] getItemForKey:itemKey];
    resolve(item);
}

@end

// Android
// MCRNStorageModule.java
package com.meicai.react.storage;

import com.facebook.react.bridge.Promise;
import com.facebook.react.bridge.ReactApplicationContext;
import com.facebook.react.bridge.ReactContextBaseJavaModule;
import com.facebook.react.bridge.ReactMethod;
import com.facebook.react.bridge.ReadableArray;

/**
 * 存储管理
```

```java
    */
public class MCRNStorageModule extends ReactContextBaseJavaModule {

    public MCRNStorageModule(ReactApplicationContext reactContext) {
        super(reactContext);
    }

    @Override
    public String getName() {
        return "MCRNStorage";
    }

    @ReactMethod
    // 存入数据
    public void setItem(String key, String value) {
        MCRNStorage.getInstance(getCurrentActivity()).setItem(key, value);
    }

    @ReactMethod
    // 获取数据
    public void getItem(String key, Promise promise) {
        promise.resolve(MCRNStorage.getInstance(getCurrentActivity()).getItem(key));
    }
}
```

3. 使用方式

这样封装好的模块就可以提供给 iOS、Android 和 JavaScript 3 端共同使用。

```objc
// iOS 端调用方式
#import "DemoNativeViewController.h"
#import <MCRNStorage.h>

@implementation DemoNativeViewController
……
[MCRNStorage setItem:item forKey:key];
……
NSString *res = [MCRNStorage getItemForKey:key];
……
@end
```

```
// Android端调用方式

import com.meicai.react.storage.MCRNStorage;
……
public class MainApplication extends Application implements ReactApplication {
    ……
    MCRNStorage.getInstance(mContext).getItem(key);
    ……
    MCRNStorage.getInstance(mContext).setItem(key, value);
    ……
}

// JavaScript端调用方式

import React, { Component } from 'react';
import { NativeModules } from 'react-native'
const { MCRNStorage } = NativeModules;

class Example extends Component {
    ……
    MCRNStorage.setItem(key, value);
    ……
    MCRNStorage.getItem(key);
    ……
}
```

5.5 本章小结

React Native官方并没有提供多少媒体及本地能力的支持，大部分插件都依赖于社区，且质量参差不齐。而随着iOS和Android自身系统的升级，这些插件也需要迭代以适应新的版本，否则兼容性就可能存在问题，例如AndroidX的升级。这对于纯粹的JavaScript开发者来说是非常劝退的事情，一旦遇到这样的场景，就要花费一定的成本去寻找满足当前业务需求的原生插件，有的时候还不一定有结果。这样其实仍然会需要原生开发者的介入，提供稳定、有效的原生能力。移动端团队的构成也很有可能变为少量原生开发者配合多数JavaScript开发者的方式，原生开发者提供原生模块的支持，页面及交互等业务场景则交于JavaScript开发者来完成。

第6章 动画

动画效果的流畅性在一定程度上会影响移动端App（应用或应用程序）的用户体验，掉帧、卡顿会给用户带来不友好的感觉。在React Native这种跨平台方案中，由于要与JavaScript端通信，开发者通常会认为React Native的动画性能较为低下。那究竟React Native是如何实现动画的，所谓的性能差异又会有多大的影响，React Native的动画机制到底能支撑到什么程度的动画细节？本章将详细介绍具体的动画实现，以及帮用户规避一些因不当操作而导致的性能问题。

6.1 布局动画——LayoutAnimation

LayoutAnimation是React Native提供的一套全局布局动画API，只需要配置好动画的相关属性（例如大小、位置、透明度），然后调用组件的setState引起重绘，这些布局变化就会在下一次渲染时以动画的形式呈现。

6.1.1 基本用法

下面用一个简单的例子介绍LayoutAnimation的基本用法。

```
import React from 'react';
import {
  NativeModules,
  LayoutAnimation,
  Text,
  TouchableOpacity,
  StyleSheet,
  View,
} from 'react-native';

const { UIManager } = NativeModules;
```

```
// 在Android设备上使用LayoutAnimation，需要通过UIManager手动启用
// 且需要放在任何动画代码之前，比如可以放在入口文件App.js中
UIManager.setLayoutAnimationEnabledExperimental
  && UIManager.setLayoutAnimationEnabledExperimental(true);

export default class App extends React.Component {
  state = {
    width: 100,
    height: 100,
  };

  onPress = () => {
    LayoutAnimation.linear();
    this.setState({
      width: this.state.width + 15,
      height: this.state.height + 15
    })
  }

  render() {
    const { width, height } = this.state;
    return (
      <View style={styles.container}>
        <View style={[styles.box, { width, height }]} />
        <TouchableOpacity onPress={this.onPress}>
          <View style={[styles.button]}>
            <Text style={styles.buttonText}>布局动画</Text>
          </View>
        </TouchableOpacity>
      </View>
    );
  }
}

const styles = StyleSheet.create({
  container: {
    flex: 1,
    alignItems: 'center',
    justifyContent: 'center',
  },
```

```
  box: {
    width: 200,
    height: 200,
    backgroundColor: 'red',
  },
  button: {
    backgroundColor: 'black',
    paddingHorizontal: 20,
    paddingVertical: 15,
    marginTop: 15,
  },
  buttonText: {
    color: '#fff',
    fontWeight: 'bold',
  },
});
```

单击按钮可以看到其中的红色视图以线性动画的方式在改变自身的宽度和高度。除了宽和高之外，视图的位置和透明度也可以改变，但透明度、大小和位置不能同时形成动画，例如：

```
export default class App extends React.Component {
  state = {
    width: 100,
    height: 100,
    top: 0,
    left: 0,
    opacity: 1,
  };

  onPress = () => {
    LayoutAnimation.linear();
    this.setState({
      width: this.state.width + 15,
      height: this.state.height + 15,
      top: this.state.top + 10,
      left: this.state.left + 10,
    })
  };

  onOpacityPress = () => {
    LayoutAnimation.linear();
```

```jsx
      this.setState({
        opacity: this.state.opacity ? 0 : 1,
      })
    }

    render() {
      const { width, height, top, left, opacity } = this.state;
      return (
        <View style={styles.container}>
          {/* 透明度变化 */}
          { opacity ? <View style={[styles.box, { opacity }]} /> : null }
          {/* 大小变化 */}
          <View style={[styles.box, { width, height }]} />
          {/* 绝对定位位置变化 */}
          <View style={[styles.absoluteBox, { width, height, top, left }]} />
          <TouchableOpacity onPress={this.onPress}>
            <View style={[styles.button]}>
              <Text style={styles.buttonText}>布局动画</Text>
            </View>
          </TouchableOpacity>
          <TouchableOpacity onPress={this.onOpacityPress}>
            <View style={[styles.button]}>
              <Text style={styles.buttonText}>透明度动画</Text>
            </View>
          </TouchableOpacity>
        </View>
      );
    }
  }
```

单击"布局动画"或"透明度动画"按钮，可以观测到LayoutAnimation的动画效果。可以看到，当opacity为0时，除了组件本身的透明度有变化，由于组件渲染状态转为null，后续兄弟组件位置的改变受到了影响，并且这个位置变化也是一个动画效果，并不是突兀的直接变化，这也就是称LayoutAnimation是一个全局的布局动画的原因。它并不只影响state作用的组件，所有在这个重绘阶段产生的组件布局的变化，都受这个配置影响。

上述这几个例子只使用了LayoutAnimation的默认配置，那LayoutAnimation具体提供了哪些属性和方法呢？下面将做个具体的解析。

LayoutAnimation源码位于Libraries/LayoutAnimation，文件本身仅100多行代码，大部分动画是利用

原生端能力实现的。从LayoutAnimation.js文件中我们可以看到，LayoutAnimation主要有以下的属性和方法。

```
const LayoutAnimation = {
  /**
   * Schedules an animation to happen on the next layout.
   *
   * @param config Specifies animation properties:
   *
   *    - 'duration' in milliseconds
   *    - 'create', 'AnimationConfig' for animating in new views
   *    - 'update', 'AnimationConfig' for animating views that have been updated
   *
   * @param onAnimationDidEnd Called when the animation finished.
   * Only supported on iOS.
   * @param onError Called on error. Only supported on iOS.
   */
  configureNext,
  /**
   * Helper for creating a config for 'configureNext'.
   */
  create,
  Types: Object.freeze({
    spring: 'spring',
    linear: 'linear',
    easeInEaseOut: 'easeInEaseOut',
    easeIn: 'easeIn',
    easeOut: 'easeOut',
    keyboard: 'keyboard',
  }),
  Properties: Object.freeze({
    opacity: 'opacity',
    scaleX: 'scaleX',
    scaleY: 'scaleY',
    scaleXY: 'scaleXY',
  }),
  checkConfig(...args: Array<mixed>) {
    console.error('LayoutAnimation.checkConfig(...) has been disabled.');
  },
  Presets,
```

```
  easeInEaseOut: configureNext.bind(null, Presets.easeInEaseOut),
  linear: configureNext.bind(null, Presets.linear),
  spring: configureNext.bind(null, Presets.spring),
};
```

从本质上讲，使用LayoutAnimation实际就是调用了LayoutAnimation.configureNext方法，linear和spring只是配置了默认值的configureNext方法。该方法接收两个参数——config和onAnimationDidEnd，其中config的数据结构如下。

```
config = {
  duration: 500,
  create: AnimationConfig,      // 组件创建时的动画参数
  update: AnimationConfig,  // 组件更新时的动画参数
  delete: AnimationConfig  // 组件销毁时的动画参数
}

AnimationConfig = {
  duration: number,      // 动画时长
  delay: number,     // 延迟时长
  springDamping: number,        // 弹性阻尼系数，配合spring使用
  initialVelocity: number,  // 初始冲量
  type: Type, // 动画方式，值包括'spring' | 'linear' | 'easeInEaseOut' | 'easeIn' | 'easeOut' | 'keyboard'
  property: Property, // 布局属性，值包括'opacity' | 'scaleX' | 'scaleY' | 'scaleXY'
}
```

所以，如果要自定义动画的话，可以将上述例子中的LayoutAnimation.linear()替换为：

```
LayoutAnimation.configureNext(
  LayoutAnimation.create(
    2000,
    LayoutAnimation.Types.easeInEaseOut,
    LayoutAnimation.Properties.opacity
  ), () => {
    console.log('Animation End')
  }
);
```

不过需要注意的是，第二个参数onAnimationDidEnd目前仅支持iOS设备。LayoutAnimation.create的逻辑很简单，仅仅是把create和delete设置为同一个属性，update则使用默认的属性。

```
function create(
  duration: number,
  type: Type,
  property: Property,
): LayoutAnimationConfig {
  return {
    duration,
    create: {type, property},
    update: {type},
    delete: {type, property},
  };
}
```

如果想完全自定义 create、update 和 delete 3 个阶段的话，也可以自己重写，例如：

```
LayoutAnimation
  .configureNext({
    duration: 2000,
    create: {
      type: LayoutAnimation.Types.linear,
      delay: 1000,
      property: LayoutAnimation.Properties.scaleX,
    },
    update: {
      initialVelocity: 1,
      springDamping: 0.5,
      type: LayoutAnimation.Types.spring,
      property: LayoutAnimation.Properties.scaleXY,
    },
    delete: {
      type: LayoutAnimation.Types.easeIn,
      property: LayoutAnimation.Properties.opacity,
    }
  })
```

也就是在创建组件时，用一个延迟 1s、时长为 2s 的线性 X 轴增长（从 0 到组件原本的宽度）渐变呈现；组件更新时，则使用 spring 弹性变化；最后组件销毁时，以透明度从 1 到 0 的方式消失，不过只要执行销毁，与之相关的其他组件布局就已经无视它的占位，会随之产生相应的位置变化。

6.1.2 原生实现原理

LayoutAnimation实现的方式自然是利用了原生端提供的视图动画的能力，iOS中UIView的animate-WithDuration、Android的android.view.animation.Animation这两种动画实现方式都是基于各自平台最基础的视图组件，所以也适用于React Native的RCTRootView和ReactRootView。

LayoutAnimation的相关类如图6.1所示。

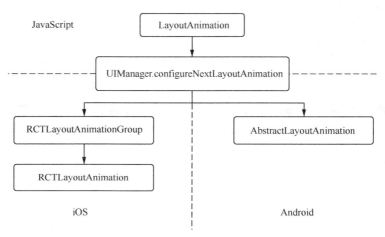

图6.1　LayoutAnimation相关类

其中的RCTLayoutAnimation和AbstractLayoutAnimation分别是iOS和Android中执行动画效果的类，接收动画作用的视图、时长、变化的视图属性等具体效果参数，例如：

```
//  iOS
//  RCTLayoutAnimation.m
……
- (void)performAnimations:(void (^)(void))animations
    withCompletionBlock:(void (^)(BOOL completed))completionBlock
{
  if (_animationType == RCTAnimationTypeSpring) {
    // iOS中可实现动画的具体效果参数
    [UIView animateWithDuration:_duration
                          delay:_delay
         usingSpringWithDamping:_springDamping
          initialSpringVelocity:_initialVelocity
                        options:UIViewAnimationOptionBeginFromCurrentState
                     animations:animations
                     completion:completionBlock];
```

```objc
  } else {
    UIViewAnimationOptions options =
      UIViewAnimationOptionBeginFromCurrentState |
      UIViewAnimationOptionsFromRCTAnimationType(_animationType);
    // iOS中可实现动画的具体效果参数
    [UIView animateWithDuration:_duration
                          delay:_delay
                        options:options
                     animations:animations
                     completion:completionBlock];
  }
}
......
```

```java
// Android
// AbstractLayoutAnimation.java
......
// 创建具体的动画类
public final @Nullable Animation createAnimation(
    View view,
    int x,
    int y,
    int width,
    int height) {
  if (!isValid()) {
    return null;
  }
  // 根据config的不同，由具体的实现类来创建最终的结果
  Animation animation = createAnimationImpl(view, x, y, width, height);
  if (animation != null) {
    int slowdownFactor = SLOWDOWN_ANIMATION_MODE ? 10 : 1;
    animation.setDuration(mDurationMs * slowdownFactor);
    animation.setStartOffset(mDelayMs * slowdownFactor);
    animation.setInterpolator(mInterpolator);
  }
  return animation;
}
......
```

调用UIManager.configureNextLayoutAnimation实际上就是创建了RCTLayoutAnimation/AbstractLayoutAnimation，并且会加入UI的更新队列，在重新Layout时调用这些动画效果。例如iOS RCTUIManager中的uiBlockWithLayoutUpdateForRootView。

```objc
......
// RCTUIManager.m

-(RCTViewManagerUIBlock)uiBlockWithLayoutUpdateForRootView:(RCTRootShadowView *)rootShadowView
{
    ......
    // 重新布局
    return ^(__unused RCTUIManager *uiManager, NSDictionary<NSNumber *, UIView *> *viewRegistry) {
        ......
        if (creatingLayoutAnimation) {
            // 创建动画
            ......
            [creatingLayoutAnimation performAnimations:^{
                if (
                    [property isEqualToString:@"scaleX"] ||
                    [property isEqualToString:@"scaleY"] ||
                    [property isEqualToString:@"scaleXY"]
                ) {
                    view.layer.transform = finalTransform;
                } else if ([property isEqualToString:@"opacity"]) {
                    view.layer.opacity = finalOpacity;
                }
            } withCompletionBlock:completion];
        } else if (updatingLayoutAnimation) {
            // 动画视图更新
            [updatingLayoutAnimation performAnimations:^{
                [view reactSetFrame:frame];
            } withCompletionBlock:completion];
        } else {
            // 无动画，直接更新视图
            [view reactSetFrame:frame];
            completion(YES);
        }
    }
}
```

```
  ……
}
```

而Android的流程会长一些，相关类如图6.2所示。

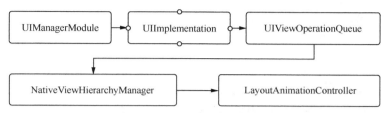

图6.2　Android动画实现相关类

无论是动画的创建还是执行，基本都遵循了这个流程：UIManagerModule发起configureNextLayout-Animation配置，创建的任务会加入UIViewOperationQueue维护的UI操作队列中；执行后会调用NativeViewHierarchyManager中的LayoutAnimationController对象初始化这个动画的配置。在视图更新后，依旧由UIManagerModule发起通知，在UIViewOperationQueue的队列中执行UpdateLayoutOperation任务，最后调用LayoutAnimationController中的applyLayoutUpdate完成动画效果，例如：

```
// LayoutAnimationController.java
public void applyLayoutUpdate(View view, int x, int y, int width, int height) {
  UiThreadUtil.assertOnUiThread();

  final int reactTag = view.getId();
  ……
  Animation animation = layoutAnimation.createAnimation(view, x, y, width, height);

  if (animation instanceof LayoutHandlingAnimation) {
    // 增加动画监听
    animation.setAnimationListener(new Animation.AnimationListener() {
      @Override
      public void onAnimationStart(Animation animation) {
        mLayoutHandlers.put(reactTag, (LayoutHandlingAnimation) animation);
      }

      @Override
      public void onAnimationEnd(Animation animation) {
        mLayoutHandlers.remove(reactTag);
      }
```

```
        @Override
        public void onAnimationRepeat(Animation animation) {}
    });
} else {
    // 修改视图位置
    view.layout(x, y, x + width, y + height);
}

if (animation != null) {
    long animationDuration = animation.getDuration();
    if (animationDuration > mMaxAnimationDuration) {
        mMaxAnimationDuration = animationDuration;
        scheduleCompletionCallback(animationDuration);
    }
    // 开始执行动画
    view.startAnimation(animation);
}
}
```

6.2 交互动画——Animated

Animated 是 React Native 提供的另一种动画方式，较之于 LayoutAnimation，它更为精细，可以只作用于单个组件的单个属性，也可以根据手势的响应来设定动画（例如通过手势放大图片等行为），甚至可以将多个动画变化组合到一起，并可以根据条件中断或修改。那能实现如此多的功能，Animated 在使用的过程中是否会过于复杂呢？本节就为大家介绍其具体的用法及一些常见的使用技巧。

6.2.1 基本用法

使用 Animated 最基本的方式就是创建一个 Animated.Value，并将它赋值到动画组件的样式属性中，再通过 Animated.timing 方法设定具体的动画效果。所谓的动画组件指可以被 Animated 封装的 React Native 组件，共计 6 个，分别是 View、Text、Image、ScrollView、FlatList 和 SectionList。所以，一个最基本的使用 Animated 的例子就可以写成：

```
import React from 'react';
import {
  Animated,
  View,
  Text,
```

```jsx
  TouchableWithoutFeedback,
} from 'react-native';

export default class App extends React.Component {

  state = {
    width: new Animated.Value(100)
  };

  count = 1;

  onPress = () => {
    this.count++;
    Animated.timing(
      this.state.width,
      {
        toValue: 100 + 20 * this.count,
        duration: 1000,
      }
    ).start();
  };

  render() {
    return (
      <View style={{ flex: 1, justifyContent: 'center', alignItems: 'center' }}>
        <TouchableWithoutFeedback onPress={this.onPress}>
          <Animated.View
            style={{ width: this.state.width, borderWidth: 1, alignItems: 'center' }}
          >
            <Text>增加宽度</Text>
          </Animated.View>
        </TouchableWithoutFeedback>
      </View>
    );
  }
}
```

总结上面这个例子可知，要使用 Animated 开发一个动画效果，最少包含下面 3 个条件。

（1）需要产生动画的组件，例如 Animated.View。

（2）关联动画的样式属性和 state 中样式的初始值，例如 this.state.width=new Animated.Value(100)。

（3）设置动画属性最后的结束值及过程描述，并最后调用start控制动画的启动。

了解了基本用法之后，下面针对这三部分内容做一个具体的分析，看看还有哪些深层次的用法。

首先是动画组件，上文也提到了React Native中提供了总共6个支持Animated的组件，分析源码（路径react-native/Libraries/Animated/src/components）可以发现这6个组件都是通过createAnimatedComponent方法对原本组件进行的封装，并且这个方法也暴露在Animated模块中，也就是说开发者可以对自定义的组件进行动画的封装，例如：

```
import React from 'react';
import {
  Animated,
  View,
  Text,
  TouchableWithoutFeedback,
} from 'react-native';

class CustomView extends React.Component {
  render() {
    return (
      <View style={this.props.style}>
        { this.props.children }
      </View>
    )
  }
}

const AnimatedCustomView = Animated.createAnimatedComponent(CustomView);

export default class App extends React.Component {

  state = {
    width: new Animated.Value(100)
  };

  count = 1;

  onPress = () => {
    this.count++;
    Animated.timing(
      this.state.width,
```

```
      {
        toValue: 100 + 20 * this.count,
        duration: 1000,
      }
    ).start();
  };

  render() {
    return (
      <View style={{ flex: 1, justifyContent: 'center', alignItems: 'center' }}>
        <AnimatedCustomView style={{ width: this.state.width, borderWidth: 1, height: 60 }} >
          <Text>自定义动画组件</Text>
        </AnimatedCustomView>
        <TouchableWithoutFeedback onPress={this.onPress}>
          <Text>增加宽度</Text>
        </TouchableWithoutFeedback>
      </View>
    );
  }
}
```

其次，与样式属性相关联的Animated.Value，除了设置初始值之外，还提供了监听函数，因此可以将上述例子修改成：

```
……
constructor(props) {
  super(props);
  const widthValue = new Animated.Value(100);
  widthValue.addListener(this.onWidthChange);
  this.state = {
    width: widthValue
  }
}

onWidthChange = (value) => {
  console.log('onWidthChange', value);
}
……
```

最后，在动画执行的过程中就可以得到以下的输出结果，如图6.3所示。

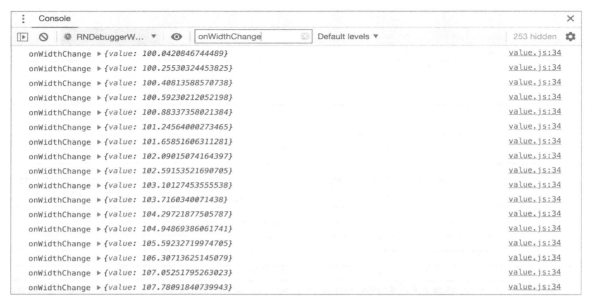

图6.3　动画执行阶段

当然，Animated既然提供了addListener方法，自然也存在removeListener以及removeAllListeners方法，例如：

```
const widthValue = new Animated.Value(100);
const animatedListenerId = this.widthValue.addListener(this.onWidthChange);
widthValue.removeListener(animatedListenerId);
// 或者
widthValue.removeAllListeners()
```

除了Animated.Value之外，Animated还提供了Animated.ValueXY()这种赋值方式，可以同时处理两个动画属性，默认key为x和y，通常用于处理组件位置动画，例如：

```
import React from 'react';
import {
  Animated,
  View,
  Text,
  TouchableWithoutFeedback,
} from 'react-native';

export default class App extends React.Component {
```

```
constructor(props) {
  super(props)
  this.state = {
    position: new Animated.ValueXY({ x: 0, y: 0 }),
  }
}

onPositionChange = () => {
  Animated.timing(
    this.state.position,
    {
      toValue: { x: 120, y: 120 },
      duration: 1000,
    }
  ).start();
}

render() {
  const { position } = this.state;
  return (
    <View style={{ flex: 1, justifyContent: 'center', alignItems: 'center' }}>
      <Animated.View
        style={[
          { position: 'absolute', borderWidth: 1, width: 50, height: 50 },
          { left: position.x, top: position.y }
        ]}
      />
      <TouchableWithoutFeedback onPress={this.onPositionChange}>
        <View>
          <Text>改变位置</Text>
        </View>
      </TouchableWithoutFeedback>
    </View>
  );
}
```

为了更方便地处理位置动画，Animated.ValueXY 提供了两个快捷方式，即 getLayout 和 getTranslate-Transform，分别将 x 和 y 转化成 left、top 和 translateX、translateY，省去手动转化的步骤，例如：

```
<Animated.View style={this.state.position.getLayout()} >
<Animated.View style={this.state.position.getTranslateTransform()} >
```

最后分析一下决定动画过程的 Animated.timing。Animated.timing 共接受如下两个参数。

- value：可以接受 AnimatedValue、AnimatedValueXY 对象，或直接是 Number 数值。
- config：配置对象，包含了常见的动画属性、回调函数及 React Native 提供的相关优化配置，具体如下。

toValue：结束值，同样可以直接是数值或 AnimatedValue（或 AnimatedValueXY）对象。

easing：缓动函数，可以通过 Easing 模块来设定，例如 Easing.inOut(Easing.ease)。

duration：动画时长，默认值为 500ms。

delay：动画开始延迟时间，默认值为 0。

isInteraction: 是否启用 Interactionmanager，它可以将一些耗时较长的工作安排到所有交互或动画完成之后再进行，以保证 JavaScript 动画的流畅运行，默认值为 true。

useNativeDriver：是否启动原生引擎，默认值为 false。使用原生引擎会改变由 JavaScript 线程计算过程值再传递给原生 UI 线程这一过程，会将动画过程直接交由原生端计算和渲染，从而减少 JavaScript 和原生通信的消耗，避免占用更多的资源而造成 UI 线程的卡顿。但并不是所有的属性都支持 useNativeDriver，在源码（路径 react-native/Libraries/Animated/src/NativeAnimatedHelper.js）中可以看到有两个白名单常量 STYLES_WHITELIST 和 TRANSFORM_WHITELIST，只有在这两个常量中的属性才可以使用 useNativeDriver。所以大部分情况下，如果要使用原生引擎改变组件的大小或位置，通常会使用 transform 属性，像 width、height、top、left、right、bottom 等这类样式则均不支持原生引擎。

onComplete: 动画结束后的回调函数，回调参数为 { finished: true }。

iterations: 动画次数，默认为 1 次，若设置为 -1 则为无限循环。

完整的例子如下。

```
Animated.timing(
  this.state.scaleX,
  {
    toValue: 2,
    duration: 1000,
    delay: 500,
    easing: Easing.inOut(Easing.ease),
    isInteraction: true,
    useNativeDriver: true,
    onComplete: (value) => { console.log('animation end', value) },
    iterations: 1,
  }
```

```
).start();
```

除了完全自定义动画过程外,Animated 模块也提供了另外两个封装好的动画函数:Animated.decay 和 Animated.spring。两者的参数除了共有参数 isInteraction/useNativeDriver/onComplete/iterations 外,其余的与 Animated.timing 略有不同,具体含义如下。

Animated.decay:以指定的初始速度开始变化,然后变化速度越来越慢直至停下。

velocity:初始速度,必填。

deceleration:衰减系数,默认值为 0.997。

Animated.spring:基于阻尼谐振(Damped Harmonic Oscillation)的弹性模型来生成动画值。它会在 toValue 值更新的同时跟踪当前的速度状态,以确保动画连贯,并且可以进行链式调用。

bounciness:弹性系数,默认值为 8;数值越大,"晃动"的幅度越大。

speed:动画速度,默认值为 12;数值越大,动画属性的变化越快。

tension:物理名词张力,默认值为 40;数值越大,"晃动"的速度越快。

friction:摩擦系数,默认值为 7;数值越大,"晃动"的次数和幅度越小。

stiffness:刚度系数,默认值为 100。刚度本身是物理名词,指材料或结构在受力时抵抗弹性变形的能力,也就是刚度系数越大,弹簧越不容易变形。所以在动画中,该值越大,"晃动"的次数越多,幅度越大,同时动画的速度也越快。

damping:阻尼系数,默认值为 10。当物体受到外力作用而振动时,会产生一种使外力衰减的反力,称为阻尼力,作用是防止物体来回抖动。所以该值越大,"晃动"的次数和幅度越小。

mass:质量,默认值为 1。物体的质量越大,惯性也就越大,所以"晃动"的幅度越大,次数越多。

velocity:初始冲量。

overshootClamping:布尔值,用于指定弹性变形过程中,动画值是否能超出 toValue 设定的值,默认值为 false,即可以超出。

restDisplacementThreshold:剩余距离阈值,默认值为 0.001。组件当前动画的值与 toValue 设定的值之差的绝对值小于该阈值,且速度也小于 restSpeedThreshold,判定为动画结束。

restSpeedThreshold:速度阈值,组件当前动画变化的速度小于该阈值,且剩余距离也小于 restDisplacementThreshold,判定为动画结束。

乍一看这几组数据都过于物理化,很难直接想象出具体的动画效果是什么样子,下面通过一个完整的例子手动调节参数来观测动画效果,以便用户选取适合自己场景的数值。

```
import React from 'react';
import {
  Animated,
  View,
  Text,
```

```
    TouchableWithoutFeedback,
    Easing,
    TextInput,
    Switch,
} from 'react-native';

export default class App extends React.Component {

    constructor(props) {
        super(props)
        this.state = {
            bounciness: 8,
            speed: 12,

            tension: 40,
            friction: 7,

            stiffness: 100,
            damping: 10,
            mass: 1,

            velocity: 0,
            overshootClamping: false,
            restDisplacementThreshold: 0.001,
            restSpeedThreshold: 0.001,

            translateX: new Animated.Value(10),
        }
    }

    onConfigChange = (value, key) => {
        this.setState({ [key]: value }, () => {
            if (['bounciness', 'speed'].indexOf(key) != -1) {
                this.springConfig = {
                    bounciness: parseFloat(this.state.bounciness),
                    speed: parseFloat(this.state.speed),
                }
            } else if (['tension', 'friction'].indexOf(key) != -1) {
                this.springConfig = {
                    tension: parseFloat(this.state.tension),
                    friction: parseFloat(this.state.friction),
                }
```

```
      } else if (['stiffness', 'damping', 'mass'].indexOf(key) != -1) {
        this.springConfig = {
          stiffness: parseFloat(this.state.stiffness),
          damping: parseFloat(this.state.damping),
          mass: parseFloat(this.state.mass),
        }
      }
    });
  }

  onStart = () => {
    this.setState({ translateX: new Animated.Value(10) }, () => {
      Animated.spring(
        this.state.translateX,
        {
          toValue: 200,
          useNativeDriver: true,
          velocity: parseFloat(this.state.velocity),
          overshootClamping: !!this.state.overshootClamping,
          restDisplacementThreshold: parseFloat(this.state.restDisplacementThreshold),
          restSpeedThreshold: parseFloat(this.state.restSpeedThreshold),
          ...this.springConfig
        }
      ).start();
    })
  };

  render() {
    const { translateX, ...restConfig } = this.state;
    return (
      <View style={{ flex: 1, justifyContent: 'center' }}>
        <Animated.View style={{ transform: [{ translateX }], width: 50, height: 50, backgroundColor: 'black' }} />
        <TouchableWithoutFeedback onPress={this.onStart}>
          <View style={{ alignItems: 'center', paddingVertical: 5 }}>
            <Text>开始动画</Text>
          </View>
        </TouchableWithoutFeedback>
        {
          Object.keys(restConfig).map(attr => (
            <View style={{ flexDirection: 'row', paddingHorizontal: 16,
```

```
paddingVertical: 5 }}>
            <Text>{ attr }: </Text>
            {
              typeof this.state[attr] === "boolean"
                ? <Switch
                    value={this.state[attr]}
                    onValueChange={value => this.setState({ [attr]: value })}
                  />
                : <TextInput
                    style={{ padding: 0, borderBottomWidth: 1, width: 100 }}
                    value={this.state[attr].toString()}
                    onChangeText={value => this.onConfigChange(value, attr)}
                  />
            }
          </View>
        ))
      }
    </View>
  );
}
```

实际效果如图6.4所示。

图6.4　动画属性效果示例

需要注意的是，不能同时定义bounciness/speed、tension/friction或stiffness/damping/mass这3组数据，只能指定其中一组来控制弹性动画。

6.2.2 动画的控制与组合

上述小节主要描述了单个属性变化的动画效果，但在实际场景中我们遇到的动画需求通常会要求处理多个属性变化，甚至需要改变多个组件的样式；又或者对动画流程要有交互的控制，例如顺序执行、中断过程等操作。那React Native是否能够处理这些更为精细的动画需求呢？本小节就来说明一下Animated提供的具体方法。

1. 动画中断及重置

Animated中的timing、spring和decay执行后都会返回一个对象，其中就包含了动画终止及重置的方法，包括以下3种。

（1）start：动画开始执行。

（2）stop：终止动画。

（3）reset：动画重置。

具体示例如下。

```
import React from 'react';
import {
  Animated,
  View,
  Text,
  TouchableWithoutFeedback,
  Easing,
  StyleSheet
} from 'react-native';

export default class App extends React.Component {

  constructor(props) {
    super(props)
    this.state = {
      translateX: new Animated.Value(-100),
    }
  }

  onPress = () => {
    this.animation = Animated.timing(
```

```
      this.state.translateX,
      {
        toValue: 200,
        duration: 10000,
        useNativeDriver: true,
        onComplete: (value) => { console.log('animation end', value) },
      }
    );
    this.animation.start();
};

onStop = () => {
    this.animation && this.animation.stop();
};

onReset = () => {
    this.animation && this.animation.reset();
}

render() {
    console.log('render', this.state.translateX)
    return (
      <View style={{ flex: 1, justifyContent: 'center', alignItems: 'center' }}>
        <Animated.View
          style={[
            { width: 50, height: 50, borderWidth: 1 },
            { transform: [{ translateX: this.state.translateX }]}
          ]}
        />
        <View style={{ flexDirection: 'row', marginTop: 40 }}>
          <TouchableWithoutFeedback onPress={this.onPress}>
            <View style={styles.button}>
              <Text>开始</Text>
            </View>
          </TouchableWithoutFeedback>
          <TouchableWithoutFeedback onPress={this.onStop}>
            <View style={styles.button}>
              <Text>结束</Text>
            </View>
          </TouchableWithoutFeedback>
```

```
        <TouchableWithoutFeedback onPress={this.onReset}>
          <View style={styles.button}>
            <Text>重置</Text>
          </View>
        </TouchableWithoutFeedback>
      </View>
    </View>
    );
  }
}

const styles = StyleSheet.create({
  button: {
    width: 50,
    height: 50
  },
});
```

2. 顺序动画

Animated 提供了 sequence 方法,接受一个动画数组并按顺序进行执行,如果当前的动画被中止,后面的动画也不会继续执行。我们可以用一个折线动画示例来展示 sequence 的用法。

```
export default class App extends React.Component {

  constructor(props) {
    super(props)
    this.state = {
      translateX: new Animated.Value(0),
      translateY: new Animated.Value(0),
    }
  }

  onPress = () => {
    Animated.sequence([
      Animated.timing(this.state.translateX, { toValue: 200, duration: 1000 }),
      Animated.timing(this.state.translateY, { toValue: 200, duration: 1000 }),
    ]).start();
  };

  render() {
```

```
      const { translateX, translateY } = this.state;
      return (
        <View style={{ flex: 1, justifyContent: 'center' }}>
          <Animated.View
            style={[
              { width: 50, height: 50, borderWidth: 1 },
              { transform: [{ translateX }, { translateY }]}
            ]}
          />
          <TouchableWithoutFeedback onPress={this.onPress}>
            <View>
              <Text>开始</Text>
            </View>
          </TouchableWithoutFeedback>
        </View>
      );
    }
  }
```

3. 并行动画

Animated.parallel 提供了同时启动多个动画的效果，接受的参数与 sequence 一致，我们只需要将上述例子中的 sequence 替换成 parallel 就可以得到一个斜线位移的动画效果，例如：

```
Animated.parallel([
  Animated.timing(this.state.translateX, { toValue: 200, duration: 1000 }),
  Animated.timing(this.state.translateY, { toValue: 200, duration: 1000 }),
]).start();
```

4. 延时动画

Animated 提供了一个单纯的延时方法——delay，通常会用在组合动画中，例如：

```
Animated.sequence([
  Animated.timing(this.state.translateX, { toValue: 200, duration: 1000 }),
  Animated.delay(500)
  Animated.timing(this.state.translateY, { toValue: 200, duration: 1000 }),
]).start();
```

这样在完成 x 轴的位移后，就会停留 500ms，然后再进行 y 轴的位移。Animated.delay 的实现方式也非常简单，就是封装了一个 timing 方法，并设置了其中的 delay 参数。

```
timing(new AnimatedValue(0), {toValue: 0, delay: time, duration: 0});
```

5. 延时并行动画

Animated.stagger 提供了一个并行动画，但每个动画会受指定的延时影响，分析它的源码就能很清楚地了解它的作用。

```
const stagger = function(
  time: number,
  animations: Array<CompositeAnimation>,
): CompositeAnimation {
  return parallel(
    animations.map((animation, i) => {
      return sequence([delay(time * i), animation]);
    }),
  );
};
```

stagger 接受两个参数，一个 time 数值，以及一个动画数组。整个动画数组都是并行开始的，只不过根据数组位置会被设置相应 time 倍数的延时。

6.2.3 动画值的运算与变化

除了直接使用 Animated.Value 创建动画值之外，Animated 模块还提供了几种运算方法，计算两个动画值得到一个新的动画值，如下所示。

Animated.add(a, b)：a + b。

Animated.subtract(a, b)：a − b。

Animated.divide(a, b)：a / b。

Animated.multiply(a, b)：a * b。

Animated.modulo(a, modulus)：取 a 除以 modulus 的余数。

Animated.diffClamp(a, min, max)：新动画值会跟随动画值 a 的变化而变化，但被限定了最小和最大区间，即使 a 值超出了这个范围，新动画值也在 min 和 max 区间。

乍一看这很大程度上只是简单的运算处理，那为什么要特意提供一些特定的方法来实现呢？其实，这些方法的本质都是创建了一个与原始动画值相关联的新动画值，这样在设定一次动画的过程中就可以同时作用于多个动画值，而不需要使用 Animated.parallel 方法。我们可以将上述并行动画的例子直接修改成：

```
export default class App extends React.Component {
```

```
constructor(props) {
  super(props)
  this.translateX = new Animated.Value(0),
  this.translateY = Animated.add(this.translateX, 0);
}

onPress = () => {
  Animated.timing(this.state.translateX, { toValue: 200, duration: 1000 }).start();
};

render() {
  const { translateX, translateY } = this;
  return (
    <View style={{ flex: 1, justifyContent: 'center' }}>
      <Animated.View
        style={[
          { width: 50, height: 50, borderWidth: 1 },
          { transform: [{ translateX }, { translateY }]}
        ]}
      />
      <TouchableWithoutFeedback onPress={this.onPress}>
        <View>
          <Text>开始</Text>
        </View>
      </TouchableWithoutFeedback>
    </View>
  );
}
```

也能得到同样的效果,但动画过程将被缩减成一个timing来控制。

除了上述这些方法外,Animated.Value还提供了一个差值转化的方式,将输入值的区间映射成另一个输出的区间,例如:

```
style={{
  opacity: this.state.fadeAnim,
  transform: [{
    translateY: this.state.fadeAnim.interpolate({
      inputRange: [0, 1],
      outputRange: [150, 0]
```

```
  }),
 }],
}}
```

这段代码表示当 fadeAnim 的值在 0 到 1 之间变化时，作用于这个组件的 translateY 的样式则在 150 到 0 之间变化，默认这一变化是线性的，也就是说当 fadeAnim 的值为 0.5 时，输出的 translateY 的值为 75。

除了单一区间的映射外，interpolate 还支持定义多区间段落、静止区间和字符串映射，例如：

```
value.interpolate({
  inputRange: [-300, -100, 0, 100, 101],
  outputRange: [300, 0, 1, 0, 0]
});
```

上述示例展示了一段多区间的映射关系，以及最后如果输入值超过 100，输出值始终为 0 的静止区间。利用 interpolate 的字符串的映射，我们可以实现颜色以及带有单位的值的动画变换，比如下面的这个旋转动画。

```
style={{
  transform: [{
    rotateX: this.animation.interpolate({
          inputRange: [0, 360],
      outputRange: ["0deg", "360deg"]
    }),
  }],
}}
```

最后，interpolate 中还有其他可使用的参数，举例如下。

easing：缓动函数，映射关系可以通过缓动函数来控制。

extrapolate：超出 input 区间的策略，可选值有 extend、identity 和 clamp。其中默认值为 extend，表示允许超出；identity，表示超出后直接返回 input 区间的最小（大）值；clamp，表示超出后返回 output 区间的最小（大）值。

extrapolateLeft：超出 input 左区间策略，值同 extrapolate。

extrapolateRight：超出 input 右区间策略，值同 extrapolate。

6.2.4　手势跟踪

Animated.event 是一个能够快速绑定用户手势或其他事件到动画值上的方法，可以通过一个结构化的映射语法来实现，例如：

```
// 其中 this.scrollX 为一个Animated.Value对象
<ScrollView
  onScroll={
    Animated.event(
      [{
        nativeEvent: {
          contentOffset: {
            x: this.scrollX
          }
        }
      }]
    )
  }
/>
```

以上方式即等价于:

```
<ScrollView onScroll={event => this.scrollX = event.nativeEvent.contentOffset.x} />
```

也就是说Animated.event接受的数组即事件参数，可以用null来忽略某一个参数，对象的key即解析的结构，我们可以再用PanResponder做个示例。

```
// this.panX和this.panY均为Animated.Value对象
<View
  onPanResponderMove={
    Animated.event(
      [
        null,
        { dx: this.panX, dy: this.panY }
      ]
    )
  }
>
  ......
</View>
```

6.3 动画实现原理及优化

在熟悉了React Native提供的动画方案之后，我们可以进一步探究这个问题——它究竟是如何利用原生能力去实现动画的？对原理的进一步了解是为了让我们能更好地使用这套动画方案，避免一些开发上的不规范、不符合框架设计本意而造成的性能问题。

6.3.1 动画实现原理

计算机动画本质上是用一个固定的频率来更新变化的图像，从而在视觉上产生"动"这个感觉。目前，大部分屏幕会以每秒60次的频率更新，而各平台、系统都会对自身的动画实现进行一些封装，开发者只需要设定动画的初始值、结束值和中间过渡的缓动函数就可以实现效果，例如Web开发中的CSS动画。同时，各平台也会直接暴露出系统级别的更新接口，以屏幕刷新的频率定时调用开发者编写的动画函数，从而达到变化的效果，例如浏览器提供的requestAnimationFrame，iOS的CADisplay-Link，Android的Choreographer，都能起到类似的作用。

在React Native中，动画属性会被逐步拆解成一个更小的单位——AnimatedNode，每个AnimatedNode都具备自己的唯一标志、属性值和监听事件等。React Native Animated 相关类如图 6.5 所示。

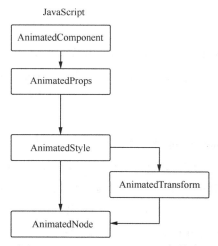

图 6.5 React Native Animated 相关类

通过Animated封装的组件会将props参数style中通过AnimatedValue封装的动画样式属性值，生成对应AnimatedStyle和AnimatedTransform的实例（这两个类的基类均为AnimatedNode）。在启动动画后，组件会通过AnimatedNode提供的＿＿getAnimatedValue方法来获取当前样式属性并进行修改，从而达到动画的效果。

JavaScript中的AnimatedNode在新建时生成一个唯一标识NodeTag，利用这个NodeTag并通过NativeAnimatedHelper.API.createAnimatedNode方法在原生端创建一个对应的NativeAnimatedNode，如图6.6所示。根据AnimatedNode设置的不同，会有以下两段两种不同的动画实现流程。

（1）由AnimatedNode计算值的变化，再通知NativeAnimatedNode修改。

（2）NativeAnimatedNode通过AnimatedNode初始化时传入的配置自己计算动画值，并通过onAnimatedValueUpdate 事件通知 AnimatedNode 同步。

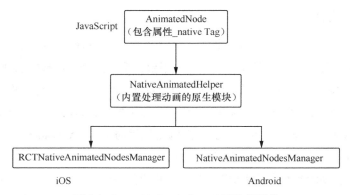

图 6.6　React Native Animated 原生相关类

6.3.2　常见优化手段

在第 4 章中我们通过 setState 重新渲染组件位置的方式实现了一个拖曳效果，但频繁调用 render 显然不是一个高效的做法，结合这一章介绍的 Animated 的模块，本小节会介绍如何开发性能更优的动画效果。

1. setNativeProps

通常情况下，我们会使用 state 和 props 来影响组件的样式属性，但这样每次的修改都会触发 render，引起组件的重绘，而过度使用 render 无疑加重了 JavaScript 线程的负担。在浏览器中，我们可以直接操作 DOM 修改样式，这样可以越过 React 提供的 diff 机制来提升性能（实际场景中并不推荐）。在移动端，React Native 也提供了类似的方法——setNativeProps，可以直接修改原生组件的样式而不触发 render，所以，我们可以将 4.3.2 小节的拖曳例子改造成如下代码段：

```
import React, { Component } from 'react';
import { PanResponder, View } from 'react-native';

export default class App extends Component {

  componentWillMount() {
    this.startX = 0;
    this.startY = 0;

    this._dragResponder = PanResponder.create({
      onStartShouldSetPanResponder: (evt, gestureState) => true,
      onPanResponderMove: (evt, gestureState) => {
```

```
          this._component.setNativeProps({
            style: {
              transform: [
                { translateX: this.startX + gestureState.dx },
                { translateY: this.startY + gestureState.dy },
              ]
            }
          })
        },
        onPanResponderEnd: (evt, gestureState) => {
          this._component.setNativeProps({
            style: {
              transform: [
                { translateX: this.startX + gestureState.dx },
                { translateY: this.startY + gestureState.dy },
              ]
            }
          })
        }
    });
  }

  render() {
    console.log('render'); // 只触发一次，拖曳过程中不会再次触发
    return (
      <View style={{ flex: 1, justifyContent: 'center', alignItems: 'center' }}>
        {/* 拖曳 */}
        <View
          ref={component => this._component = component}
          style={{{
            width: 100,
            height: 50,
            borderWidth: 1,
          }}
          {...this._dragResponder.panHandlers}
        />
      </View>
    )
  }
}
```

在 Animated 模块中，我们也可以追溯到源码的 react-native/Libraries/Animated/src/createAnimated-Component 中的 _animatedPropsCallback，这里也使用了 this._component.setNativeProps 修改组件的样式，也就是说使用 Animated 模块开发的动画（非 useNativeDriver 模式下）会直接修改原生组件的样式，而不会触发 render 重绘，从而减少 JavaScript 线程的工作。

2. InteractionManager

Interactionmanager 模块提供了一些方法，可以将一些耗时较长的工作安排到所有交互或动画完成之后再进行，以确保动画不被阻塞，例如：

```
InteractionManager.runAfterInteractions(() => {
  // 可以放置执行耗时较长的同步任务，例如执行 setState 导致的重绘大量组件
});
```

并且可以通过 InteractionManager.createInteractionHandle 在动画开始时创建一个句柄，然后在结束的时候清除这个句柄，例如：

```
const handle = InteractionManager.createInteractionHandle();
// 执行动画... ('runAfterInteractions' 中的任务现在开始排队等候)
// 在动画完成之后开始清除句柄
InteractionManager.clearInteractionHandle(handle);
// 在所有句柄都清除之后，开始依序执行队列中的任务
```

在 Animated 的 config 参数中也有 isInteraction 属性，默认也使用了这种方式来避免动画的阻塞，我们可以在源码 react-native/Libraries/Animated/src/nodes/AnimatedValue.js 中看到这一使用方式。

```
animate(animation: Animation, callback: ?EndCallback): void {
  let handle = null;
  // 在动画开始前创建句柄
  if (animation.__isInteraction) {
    handle = InteractionManager.createInteractionHandle();
  }
  const previousAnimation = this._animation;
  this._animation && this._animation.stop();
  this._animation = animation;
  animation.start(
    this._value,
    value => {
      // Natively driven animations will never call into that callback,
therefore we can always
      // pass flush = true to allow the updated value to propagate to native
```

```
with setNativeProps
        this._updateValue(value, true /* flush */);
      },
      result => {
        this._animation = null;
        // 动画结束后清除句柄
        if (handle !== null) {
          InteractionManager.clearInteractionHandle(handle);
        }
        callback && callback(result);
      },
      previousAnimation,
      this,
    );
  }
```

3. useNativeDriver

在 React Native 项目早期，动画的定时和计算流程都由 JavaScript 线程负责，整体的流程基本遵循图6.7所示。

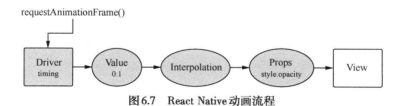

图6.7　React Native 动画流程

requestAnimationFrame 是 JavaScript 端的计时器，每帧刷新时都会调用这个函数，计算出当前时刻组件的动画值，再通过 setNativeProps 的方式传递给原生视图——iOS 的 UIView 或 Android 的 android.View。这一过程就会加剧 JavaScript 线程的计算压力，更不用说还需负责将计算结果通知给原生端。一旦 JavaScript 线程由于其他原因发生阻塞，那势必会引起动画效果的卡顿。

为了解决这个问题，React Native 之后将计时和计算过程都交给了原生端实现，也就是上文提到使用 AnimatedValue 时的 config 中的参数 useNativeDriver。将 useNativeDriver 设置为 true 后，JavaScript 端仅需在动画开始时将相关数据传给原生端，之后相关的计算和渲染将由原生端自己处理；如果你设置了属性变化的事件监听的话，也将由原生端主动发起传递给 JavaScript 端。这样也就避免了每帧的计算和更新都需要 JavaScript 线程处理，而都直接交给原生端 UI 主线程处理。

但这个方案也不是十全十美的，因为并不是所有的样式属性都支持 useNativeDriver 属性，目前的版本情况下仅支持以下几种：

- opacity；
- transform；
- borderRadius 及相关属性，例如 borderBottomEndRadius、borderTopLeftRadius 等；
- elevation；
- shadowOpacity、shadowRadius（仅 iOS 下支持）；
- transform 相关属性，translate、scale、rotate、perspective。

若使用 Interpolation 的话，仅支持 inputRange、outputRange、extrapolate、extrapolateRight 和 extrapolateLeft 属性，也就是说不支持 JavaScript 设定的缓动函数。

6.4 本章小结

在 React Native 的设计结构下，如果单纯以一定频率调用 setState 的方式去修改组件的 style 以实现动画效果，那必然会造成大量的 render 重绘及 JavaScript 与原生端的通信，从而导致 UI 线程的卡顿。因此，React Native 提供的动画模块及方法基本是围绕减少这两个方面的触发频率来实现，特别是使用 useNativeDriver，需要原生端各自实现动画的计算过程以规避与 JavaScript 之间的通信，从而实现流程的动画效果。

第7章 React Native 与原生端的通信方式

在前面几章我们已经或多或少地了解到了在 React Native 项目中，JavaScript 与原生端之间会存在着大量的通信，实现了 UI 渲染、事件通信，以及动画信号等功能。这背后引申出来的问题就是，JavaScript 究竟是如何与 Objective-C(C++)/Java 语言进行数据共享及函数调用的，React Native 又提供了什么样的方式可以使开发者新增自定义原生方法供 JavaScript 端调用呢？本章就会向大家详细介绍在 React Native 中 JavaScript 与原生端的几种通信方式。

7.1 JavaScript 调用原生模块

在 React Native 中，原生端通常以模块的形式暴露给 JavaScript 端，并声明可被调用的常量和方法，一个原生 App（应用或应用程序）中可以包含多个原生模块，每个原生模块也可以包含多个原生方法。同时，在 JavaScript 端可以使用 React Native 提供的 NativeModules 获取暴露出的原生模块，进而可以获取到其中的常量和方法。鉴于语言和平台本身的差异性，我们将分开来描述 iOS 和 Android 是如何来实现这一功能的。本节会以 5.4.4 小节涉及的混合存储方案 MCRNStorage 和一个自定义事件通信模块 MCRNEventEmitter 作为具体的示例。全部代码都包含在 //TODO。

7.1.1 iOS 与 JavaScript 的通信方式

在 iOS 中导出一个可供 React Naive JavaScript 端调用的模块，首先需要实现一个遵守 RCTBridgeModule 协议的类，并实现该协议中声明的方法。为了方便开发，React Native 提供了一个宏函数——RCT_EXPORT_MODULE，只需把该方法写在类的 @implementation（标识当前类导出为可供 React Native 调用的模块）即可，例如：

```
@interface MCRNStorage : NSObject<RCTBridgeModule>
@end

@implementation MCRNStorage
```

```
RCT_EXPORT_MODULE();

@end
```

上述这段代码便声明了一个名为MCRNStorage的类,导出的原生模块的名字默认就是当前的类名,JavaScript端可以通过这个模块名称获取到这个原生模块,例如:

```
import { NativeModules } from 'react-native';
const { MCRNStorage } = NativeModules;
```

另外,RCT_EXPORT_MODULE还支持宏接受一个参数,用于指定原生模块的名称。

```
@implementation Foo
RCT_EXPORT_MODULE(MCRNStorage);
@end

import { NativeModules } from 'react-native';
const { MCRNStorage } = NativeModules;
```

1. 原生常量

在原生模块中,开发者可以通过constantsToExport直接导出一些常量,这些常量在JavaScript端可随时且同步访问,例如:

```
……
@implementation MCRNEventEmitter

RCT_EXPORT_MODULE()
……
- (NSDictionary *)constantsToExport
{
  return @{ @"COMPONENT_DID_APPEAR": @"componentDidAppear" };
}
……
```

2. 原生方法

同声明原生模块类似,声明可供JavaScript调用的方法需要使用React Native提供的另一个宏函数——RCT_EXPORT_METHOD,例如:

```
// iOS
@implementation MCRNStorage
……
```

```
RCT_EXPORT_METHOD(clearAllData){
    // do someting
}
@end
......

// JavaScript
import { NativeModules } from 'react-native';
const { MCRNStorage } = NativeModules;

// 发起调用
MCRNStorage.clearAllData()
```

上述这个例子仅是一个最基本的调用功能,并没有实际的意义。要想让 JavaScript 端获取原生端提供的能力,在声明的方法中我们一定会关注方法的参数是什么,JavaScript 又是如何获取返回值的。

参数方面,由于语言类型的不同,我们需要将 JavaScript 与 Objective-C 的数据类型进行转化,并且遵循表 7.1 所示的映射关系。

表7.1

JavaScript	iOS
string	NSString
number	NSInteger, float, double, CGFloat, NSNumber
boolean	BOOL, NSNumber
array	NSArray,可包含本列表中的任意类型,也就是说不限定数组中只能存在一种数据类型
object	NSDictionary,可包含 string 类型的键和本列表中其他任意类型的值
function	RCTResponseSenderBlock

也就是说我们在传参时需要明确定义参数的数据类型,例如:

```
// iOS
@implementation MCRNStorage
......
RCT_EXPORT_METHOD(setItem:(NSString *)itemKey andItem:(NSString *)item){
    // do something
}
@end

// JavaScript
```

```
import { NativeModules } from 'react-native';
const { MCRNStorage } = NativeModules;

MCRNStorage.setItem('userName', 'Eloy');
```

由于跨端通信本身是异步操作,无法直接在原生方法中使用return来让JavaScript获取返回值,因此,React Native提供了回调函数和Promise两种方式帮助JavaScript端获取经过调用原生方法后得到的结果,例如:

```
// iOS
@implementation MCRNStorage
......
RCT_EXPORT_METHOD(getItem:(NSString *)
             itemKey callback:(RCTResponseSenderBlock)callback)
{
  callback(@[[NSNull null], @"result"]);
}
......
@end

// JavaScript
......
MCRNStorage.getItem('userInfo', (error, result) => {
  if (error) {
    console.error(error);
  } else {
    console.log(result); // result
  }
});
```

其中,RCTResponseSenderBlock只接受一个参数,也就是返回给JavaScript回调函数的参数数组,JavaScript端也无须对数据类型有额外的考虑,只是按顺序接受成多个参数。

Promise写法是JavaScript ES6的新特性,提供了resolve和reject执行回调,这样在JavaScript端你可以通过.then()的方式处理返回值,也可以通过async/await的方式来避免使用回调函数的写法。具体示例如下。

```
// iOS
RCT_EXPORT_METHOD(getItem:(NSString *)itemKey
             resolver:(RCTPromiseResolveBlock)resolve
             reject:(RCTPromiseRejectBlock)reject)
```

```
{
  NSString *item = ……
  if (item) {
    resolve(item);
  } else {
    NSError *error = …
    reject(@"No Data", @"No corresponding data", error);
  }
}

// JavaScript
import { NativeModules } from 'react-native'
const { MCRNStorage } = NativeModules;
MCRNStorage.getItem(key).then(value => toJSON(value))
```

需要注意的是，无论是RCTResponseSenderBlock的callback，还是RCTPromiseResolveBlock/RCTPromiseRejectBlock的resolve和reject，均只可被调用一次，否则会抛出异常。但这些方法调用的时机可以由开发者自己决定，也就是说，当JavaScript端发起原生方法的时候，原生端不一定需要立即响应，可以等待某些操作（例如选择图片、接收通知等场景）完成后再触发JavaScript端的逻辑。

3. 多线程

多线程对于JavaScript开发者来说是个略微陌生的场景，因为JavaScript语言的一大特点就是单线程，可避免在浏览器环境中多个线程同时操作DOM而带来的复杂同步问题。但在原生开发中，多线程场景则主要出现在将消耗过多资源的场景中，开发者可以将部分耗时的行为交由子线程处理，避免阻塞主线程UI渲染而造成卡顿。React Native的原生模块通常不会对自己被调用时所处的线程做任何假设，所以如果在这里执行必须主线程才能调用的API可能会存在未知的风险。同时，React Native也预留了方法，能够指定该模块运行的线程队列，例如：

```
- (dispatch_queue_t)methodQueue
{
  return dispatch_get_main_queue(); // 指定在主线程操作
}
```

对于执行时间较长的操作，我们也可以单独创建一个队列，这样就不会阻塞React Native本身的消息队列。

```
// 指定methodQueue会影响当前模块中所有的方法，这些方法均会在新队列中被执行
- (dispatch_queue_t)methodQueue
{
```

```
    return dispatch_queue_create("com.yunshanmeicai.mcrn", DISPATCH_QUEUE_SERIAL);
}
```

或者仅某个方法执行时想创建单独的队列,那可以采用这种形式:

```
RCT_EXPORT_METHOD(doSomethingExpensive:(NSString *)param
                                    callback:(RCTResponseSenderBlock)callback)
{
    // dispatch_get_global_queue开启了一个新的队列,且第一个参数用来标识任务执行的优先级,
第二个参数则是Apple保留的参数,以备将来使用,推荐固定传入0
    (dispatch_get_global_queue(DISPATCH_QUEUE_PRIORITY_DEFAULT, 0), ^{
        // 在这里执行长时间的操作
        ……
        // 你可以在任何线程/队列中执行回调函数
        callback(@[...]);
    });
}
```

4. 监听与推送

React Native 提供了一套事件监听及分发机制,除了可以从 JavaScript 端主动向原生端发起通信等待响应外,还可以从原生端主动向 JavaScript 端推送信息。首先,需要一个继承于 RCTEventEmitter 的原生模块,并在其中使用 supportedEvents 定义可被监听的事件名称,例如:

```
// MCRNEventEmitter.h
#import <React/RCTBridgeModule.h>
#import <React/RCTEventEmitter.h>

@interface MCRNEventEmitter: RCTEventEmitter<RCTBridgeModule>

@end

// MCRNEventEmitter.m
@implementation MCRNEventEmitter

RCT_EXPORT_MODULE()
……
- (NSArray<NSString *> *)supportedEvents {
    return [@"componentDidAppear", @"componentDidDisappear"];
}
……
@end
```

其次，在JavaScript端使用NativeEventEmitter模块来建立对应的监听：

```
import { NativeModules, NativeEventEmitter } from 'react-native'
const { MCRNEventEmitter } = NativeModules;

const emitter = new NativeEventEmitter(MCRNEventEmitter);

emitter.addListener('componentDidAppear', function(data) {
  console.log(data)
})
```

最后，在原生模块中使用RCTEventEmitter提供的sendEventWithName来发送数据：

```
......
- (void)didReceiveNoti:(NSNotification *)noti
{
    [self sendEventWithName:@"componentDidAppear" body:@{@"message": "message from iOS"}];
}
......
```

从上述例子可以看到，想要发送信息就必须使用这个原生模块中的方法，但如果触发这个推送的行为并不包含在这个原生模块中，那该如何通知JavaScript端呢？iOS原生的NSNotification就提供了这么一个解耦的机制。NSNotification是Apple提供的通知分派机制，可将信息广播到注册的观察者，并且观察者和发送者无须耦合关系，只需通过定义相同的消息名称就可以完成消息派发和监听。

下面的例子重写了MCRNEventEmitter模块的初始化方法，并且注册了name为kMCRNEventInternalCommunicationKey的消息监听，当接收到对应通知时会调用didReceiveNoti:方法，并在didReceiveNoti:方法中向JavaScript端发起消息推送。

```
- (instancetype)init {
    self = [super init];
    if (self) {
        [[NSNotificationCenter defaultCenter] addObserver:self selector:@selector(didReceiveNoti:) name:kMCRNEventInternalCommunicationKey object:nil];
    }
    return self;
}

- (void)didReceiveNoti:(NSNotification *)noti {
    if ([noti.object isKindOfClass:[NSDictionary class]]) {
        NSDictionary *msgDict = (NSDictionary *)noti.object;
```

```objc
        [self sendEventWithName:msgDict[@"eventName"] body:msgDict[@"body"]];
    }
}

// App内任一模块发起通知
NSDictionary *params = @{
  @"eventName": @"componentDidAppear",
  @"body": @"body"
};

[[NSNotificationCenter defaultCenter] postNotificationName:kMCRNEventInternalCommunicationKey object: params];
```

7.1.2 Android与JavaScript的通信方式

在Android系统中开发原生模块的方式比iOS会略微烦琐一些,涉及的概念和类比较多,也没有像iOS那样可以使用宏函数的方式快速生成一个原生模块。在Android上,先新建一个继承于ReactContextBaseJavaModule的类,在其中声明这个原生模块可被JavaScript使用的名称及函数,然后将这个类实例化并放置到一个ReactPackage的实现类中,最后设置将这个实现类加入ReactNativeHost之后才可以被JavaScript端使用,具体的代码如下。

```java
// 自定义的原生模块
public class MCRNStorageModule extends ReactContextBaseJavaModule {
    public MCRNStorageModule(@Nonnull ReactApplicationContext reactContext) {
        super(reactContext);
    }

    // JavaScript端可以用的模块名称
    @Nonnull
    @Override
    public String getName() {
        return "MCRNStorage";
    }
}

// ReactPackage,可以包含多个原生模块,以及自定义UI组件(下一章详解)
public class MCRNStoragePackage implements ReactPackage {
    @Nonnull
    @Override
```

```java
    public List<NativeModule> createNativeModules(@Nonnull ReactApplicationContext reactContext) {
        List<NativeModule> list = new ArrayList<>(1);
        // 将原生模块实例化, 放置到数组中
        list.add(new MCRNStorageModule(reactContext));
        return list;
    }


    // 可声明自定义UI组件, 同样放置到数组中, 本章暂不详解, 故设置为空
    @Nonnull
    @Override
    public List<ViewManager> createViewManagers(@Nonnull ReactApplicationContext reactContext) {
        return Collections.EMPTY_LIST;
    }
}


// Android启动时的Application对象
public class MainApplication extends Application implements ReactApplication {
    @Override
    public ReactNativeHost getReactNativeHost() {
        return reactNativeHost;
    }

    private final ReactNativeHost reactNativeHost = new ReactNativeHost(this) {
        @Override
        public boolean getUseDeveloperSupport() {
            return BuildConfig.DEBUG;
        }
        // 加入原生ReactPackage, 直到这一步才能确保JavaScript端可以调用刚刚声明的原生模块
        @Override
        protected List<ReactPackage> getPackages() {
            return Arrays.asList(new MCRNStoragePackage());
        }

        @Override
        protected String getJSMainModuleName() {
            return "index";
        }
    };
}
```

1. 原生常量

在ReactContextBaseJavaModule模块中重写getConstants方法即可导出常量,例如:

```java
public class MCReactJavaEventEmitterModule extends ReactContextBaseJavaModule {
……
    @Override
    public Map<String, Object> getConstants() {
        Map<String, Object> constants = new HashMap<>();
        constants.put("COMPONENT_DID_APPEAR", MCEventEmitter.ComponentDidAppear);
        constants.put("COMPONENT_DID_DISAPPEAR", MCEventEmitter.ComponentDidDisappear);
        return constants;
    }
……
```

2. 原生方法

在Android上,React Native提供了一种注解的写法来暴露方法供JavaScript端调用,例如:

```java
// Android
@ReactMethod
public void clearAllData() {
    Toast.makeText(getCurrentActivity(), "数据清理完毕", Toast.LENGTH_SHORT).show();
}
// JavaScript
import { NativeModules } from 'react-native';
const { MCRNStorage } = NativeModules;
// 发起调用
MCRNStorage.clearAllData()
```

JavaScript与Java数据类型的转化关系则如表7.2所示。

表7.2

JavaScript	Java
string	String
number	Integer, Double, Float, CGFloat, NSNumber
boolean	Boolean
array	ReadableArray, WritableArray
object	ReadableMap, WritableMap
function	Callback

所以，如果要设定带参数的函数，则需要：

```
//Android
@ReactMethod
public void setItem(String key, String value) {
    map.put(key, value);
}
// JavaScript
import { NativeModules } from 'react-native';
const { MCRNStorage } = NativeModules;
// 发起调用
MCRNStorage.setItem('userName', '张三');
```

在使用回调函数和Promise的方式上，Android也与iOS大同小异。

```
//   回调函数Callback用法
@ReactMethod
public void getItem(String key, Callback callback) {
    Log.e("mc", "getItem: ---->callback");
    if (map.containsKey(key)) {
        callback.invoke(map.get(key), map.get(key));
    } else {
        callback.invoke(null, "数据不存在");
    }
}
// JavaScript
import { NativeModules } from 'react-native';
const { MCRNStorage } = NativeModules;
// 发起调用
MCRNStorage.getItem('userName', (error, result) => {
  if (error) {
    console.error(error);
  } else {
    console.log(result); // result
  }
});

// Promise用法
@ReactMethod
public void getItem(String key, Promise promise) {
    Log.e("mc", "getItem: --------->promise");
```

```java
        if (map.containsKey(key)) {
            promise.resolve(map.get(key));
        } else {
            promise.resolve("数据不存在");
        }
    }
}

// JavaScript
import { NativeModules } from 'react-native';
const { MCRNStorage } = NativeModules;
// 发起调用
MCRNStorage.getItem('userName').then(value => this.toJSON(value))
```

需要注意的是在Android上，接收JavaScript的数组、对象类型参数和发送给JavaScript的数组、对象类型参数并不是同一个。ReadableMap、ReadableArray用于接收JavaScript传递的Object、Array，而WritableMap、WritableArray则为原生端发送JavaScript的数据类型，例如：

```java
/**
 * @param readableArray  对应React Native的Array
 * @param readableMap    对应React Native的Object
 * @param callback       对应React Native的function
 */
@ReactMethod
public void exampleMethod(ReadableArray readableArray, ReadableMap readableMap, Callback callback) {
    ArrayList<Object> arrayList = readableArray.toArrayList();
    HashMap<String, Object> map = readableMap.toHashMap();
    WritableArray writableArray = Arguments.createArray();
    for (Object object : arrayList) {
        writableArray.pushString(object + "");
    }
    WritableMap writableMap = Arguments.createMap();
    for (String key : map.keySet()) {
        writableMap.putString(key, map.get(key) + "");
    }
    callback.invoke(writableArray, writableMap);
}
```

3. 多线程

上文提到了Android中存在主线程和子线程，且通常情况下应尽量避免在主线程执行较为耗时的操作。为了解决线程切换的问题，Android 中封装了一个轻量级的异步类 AsyncTask。AsyncTask 可以在线程池中执行异步任务，并将执行进度和执行结果传递给主线程。下面以文件下载为例简单说明一下 AsyncTask 的用法。同样，你可以在 React Native 的自定义模块中使用这个异步类。

```java
public class DownloadActivity extends AppCompatActivity implements View.OnClickListener {
    private Button btnDownload;
    private ProgressBar progressBar;
    @Override
    protected void onCreate(Bundle savedInstanceState) {
        super.onCreate(savedInstanceState);
        setContentView(R.layout.activity_download);
        btnDownload = findViewById(R.id.btnDownload);
        progressBar = findViewById(R.id.progressBar);
        btnDownload.setOnClickListener(this);
    }
    @Override
    public void onClick(View v) {
        switch (v.getId()) {
            case R.id.btnDownload:
                // 初始化下载任务
                DownloadTask d1 = new DownloadTask();
                d1.executeOnExecutor(AsyncTask.THREAD_POOL_EXECUTOR, 1);
                btnDownload.setText("正在下载");
                btnDownload.setEnabled(false);
                break;
            default:
                break;
        }
    }
}
// 实现下载文件异步类
class DownloadTask extends AsyncTask<Integer, Integer, String> {
    // 后台执行方法
    @Override
    protected String doInBackground(Integer... integers) {
        for (int i = 0; i < 100; i++) {
            try {
                Thread.sleep(200L);
            } catch (InterruptedException e) {
                e.printStackTrace();
```

```java
            }
            publishProgress(i);
        }
        return null;
    }
    @Override
    protected void onProgressUpdate(Integer... values) {
        super.onProgressUpdate(values);
        progressBar.setProgress(values[0]);
        btnDownload.setText("下载中" + values[0] + "%");
    }
    @Override
    protected void onPostExecute(String s) {
        super.onPostExecute(s);
        btnDownload.setText("下载完成");
        btnDownload.setEnabled(true);
    }
    @Override
    protected void onCancelled() {
        super.onCancelled();
        btnDownload.setText("下载失败");
        btnDownload.setEnabled(true);
    }
}
```

4. 监听与推送

监听与推送在JavaScript端的使用方式和在iOS上一模一样，下面就不再赘述了，直接说明在Android中监听模块该如何实现。

```java
    public class MCReactJavaEventEmitterModule
            extends ReactContextBaseJavaModule {
        private ReactApplicationContext reactApplicationContext;
        public MCReactJavaEventEmitterModule(@Nonnull ReactApplicationContext reactContext) {
            super(reactContext);
            reactApplicationContext = reactContext;
        }
        @Override
        public Map<String, Object> getConstants() {
```

```java
        Map<String, Object> constants = new HashMap<>();
        constants.put("COMPONENT_DID_APPEAR", MCEventEmitter.ComponentDidAppear);
        constants.put("COMPONENT_DID_DISAPPEAR", MCEventEmitter.ComponentDidDisappear);
        return constants;
    }
    @Nonnull
    @Override
    public String getName() {
        return "MCRNEventEmitter";
    }
    @ReactMethod
    public void sendEventWithName() {
        reactApplicationContext.getJSModule(DeviceEventManagerModule.RCTDeviceEventEmitter.class)
                .emit(MCEventEmitter.ComponentDidAppear, "msg from Android");
    }
}
```

同iOS类似，Android也有事件广播的机制，Broadcast就是组件之间传输数据的一种机制。Android的广播也分为发送方和接收方，发送需要将发送的消息和过滤的信息装入一个Intent对象，然后通过context.sendBroadcast方法把Intent对象以广播的形式发出去，已注册的接收方会检查注册时的IntentFilter是否与发送方的Intent匹配，如果匹配则会调用BroadcastReceiver的onReceiver方法。

```java
// 发送广播
Intent intent = new Intent();
intent.setAction(ACTION);
intent.putExtra("name", "张三" + System.currentTimeMillis());
sendBroadcast(intent);
// 接收广播
class MyBroadcastReceiver extends BroadcastReceiver {
    @Override
    public void onReceive(Context context, Intent intent) {
        Log.e(TAG, "接收到广播，接收到的数据为：" + intent.getStringExtra("name"));
    }
}
```

5. Java直接调用JavaScript方法

我们可以在JavaScript端定义一个CustomJSModule（任意名称）模块，包含一个echo方法，用于接收 Java 端传递过来的一个字符串，并对这个字符串进行一定的处理。由于 JavaScript 方法无法直接回传数据给 Java 端，因此通过原生模块 CustomJavaModule 的 handleJSReturnValue 方法完成数据的回传。以下是具体的实现步骤。

（1）首先定义模块。

```javascript
// JavaScript 自定义模块
const CustomJSModule = {
  echo: function(strFromJava) {
    const msg = 'JS 收到来自 Java 的字符串：' + strFromJava;
    ......
  },
};

export default CustomJSModule;
```

（2）完成注册。

```javascript
import CustomJSModule from './examples/7.2/CustomJSModule';
const BatchedBridge = require('react-native/Libraries/BatchedBridge/BatchedBridge');

BatchedBridge.registerCallableModule(
  'CustomJSModule',
  CustomJSModule,
);
```

（3）在原生端定义同名的 Java 接口，然后通过 CatalystInstance 的 getJSModule 方法获取 JavaScript 模块实例，完成 JavaScript 方法的调用。

```java
public class TestBridgeActivity extends AppCompatActivity {

    // 继承 JavaScriptModule 接口，名称与 JavaScript 端注册的模块名称保持一致
    private interface CustomJSModule extends JavaScriptModule {
        // 定义抽象方法，具体实现在 JavaScript 端
        void echo(String strFromJava);
    }

    @Override
    protected void onCreate(Bundle savedInstanceState) {
        ......
        // 通过 CatalystInstance 的 getJSModule 方法获取 JavaScript 模块实例，完成方法调用
```

```
        currentReactContext.getCatalystInstance().getJSModule(CustomJSModule.
class).echo("Msg from Java");
    }

}
```

7.2 JavaScript 跨平台运行原理

对于 JavaScript 开发者而言，浏览器其实是一个天然的跨平台容器，我们并不会在意 JavaScript 代码是运行在 Windows 系统、macOS 系统还是 Linux 系统。随着 V8 引擎和 Node.js 的出现，JavaScript 代码逐渐跳出了浏览器这唯一的环境，开始在服务端发挥自己的特性。我们也逐渐意识到 JavaScript 代码与浏览器并不是天然的绑定关系，而是依赖各种 JavaScript 引擎对代码进行解释及运行。所以，在如今的各种移动端设备中，面对 iOS、Android 这样不同的操作系统和语言类型，又有哪些 JavaScript 引擎给我们搭建起了跨平台的桥梁呢？本节就和大家介绍一下这些运行在移动端设备中的 JavaScript 引擎及其大致的运行方式。

7.2.1 JavaScriptCore——iOS 的 JavaScript 引擎

JavaScriptCore 最先是 Apple Safari 浏览器的 JavaScript 引擎，从 iOS 7.0 之后，开发者可以在 iOS 设备上直接使用这个引擎，使 Objective-C 和 JavaScript 代码之间的交互变得更加简单和方便，其中最直接的就是以下两种应用方式。

（1）在 Objective-C 环境中执行 JavaScript 代码。

（2）将 Objective-C 对象注入 JavaScript 环境中。

下面先看一下头文件 JavaScriptCore.h。

```
#ifndef JavaScriptCore_h
#define JavaScriptCore_h

#include <JavaScriptCore/JavaScript.h>
#include <JavaScriptCore/JSStringRefCF.h>

#if defined(__OBJC__) && JSC_OBJC_API_ENABLED

#import "JSContext.h"
#import "JSValue.h"
#import "JSManagedValue.h"
#import "JSVirtualMachine.h"
```

```
#import "JSExport.h"

#endif

#endif /* JavaScriptCore_h */
```

这里已经很清晰地列出了JavaScriptCore的主要几个类，下面将逐个说明它们的具体作用。

1. JSVirtualMachine

JSVirtualMachine的字面意思为JavaScript的虚拟机，一个JSVirtualMachine的实例就是一个完整独立的JavaScript执行环境，为JavaScript的执行提供底层资源，主要用于处理并发的JavaScript执行以及处理JavaScript和Objective-C桥接对象的内存管理。需要注意的是，每个JSVirtualMachine是完全独立且完整的，拥有自身的空间和垃圾回收器，不支持将一个JSVirtualMachine中创建的值传给另一个JSVirtualMachine。

2. JSContext

JSContext即为JavaScript运行的上下文，也就是其执行环境。每一个JSContext对象都归属于一个JSVirtualMachine，且一个JSVirtualMachine中可以包含多个不同的JSContext。Objective-C可以通过JSContext来执行、访问JavaScript代码，JavaScript也可以通过JSContext来访问Objective-C的代码。

通过头文件JSContext.h，我们可以大致了解一下JSContext常用的API。

init：初始化一个JSContext，同时会创建一个新的JSVirtualMachine。

initWithVirtualMachine：在指定的JSVirtualMachine上创建JSContext。

evaluateScript：执行一段JavaScript代码，返回最后生成的一个值；也可以接受一个sourceURL的参数，并将sourceURL认作其源码URL（仅作标记用）。

currentContext：获取当前执行的JavaScript代码的context。

currentCallee：获取当前执行的JavaScript的function。

有了这些API，就可以写一段在Objective-C中与JavaScript的交互。

```
JSContext *context = [[JSContext alloc] init];
[context evaluateScript:@"var foo = { bar: 'hello react native' }"];
JSValue *foo = context[@"foo"];
NSLog(@"%@", [foo toDictionary]);    // { bar = hello react native; }
```

可以将JSContext理解为一个JavaScript代码运行的作用域，在context中通过变量名获取这个context中的变量。

3. JSValue

一个JSValue实例就是一个JavaScript值的引用，处理了JavaScript和Objective-C代码之间基本数据类型的转化（如上文所举），例如：

```
JSContext *context = [[JSContext alloc] init];
[context evaluateScript:@"var jack = {name:'Jack', age:18, car:'BMW'}"];

// JavaScript -> Native 从context获取jack变量
JSValue *jack = context[@"jack"];
NSLog(@"name=%@, age=%@, car=%@", jack[@"name"], jack[@"age"], jack[@"car"]);
// 打印：name=Jack, age=18, car=BMW

// 在原生端给Jack换个车
jack[@"car"] = @"Mercedes";
// Native -> JavaScript Jack赋值给JavaScript端
context[@"jack"] =jack ;

// 为context添加一个log函数，方便打印调试
context[@"log"] = ^(NSString *log) {
  NSLog(@"%@", log);
};

[context evaluateScript:@"log('name:'+jack.name+' age:'+jack.age+' car:'+jack.car)"];
// 打印：name:Jack age:18 car:Mercedes
```

4. JSManagedValue

一个JSManagedValue的对象主要用来包装一个JSValue对象，使之能实现自动内存管理，基本的用法就是包装要导出到JavaScript的JSValue对象，避免造成循环引用，无法销毁JSContext（因为JSValue会强引用JSContext对象）。下面先看一个错误的示例。

```
JSContext *context = [JSContext new];
[context evaluateScript:@"var jack = {name:'Jack', age:18, car:'BMW'}"];

JSValue *jack = context[@"jack"];
context[@"nativeFunction"] = ^(){
  NSLog(@"%@", jack);
};
```

上面的示例先从JSContext中获取JSValue，JSValue强引用JSContext对象，我们又为JSContext添加了一个原生方法（Block），JSContext对象会对原生方法强引用。原生方法的方法体又使用到了JSValue对象，形成循环引用。

为了解决循环引用导致的内存泄漏，下面使用JSManagedValue来解决这个问题。

```objectivec
JSContext *context = [JSContext new];
[context evaluateScript:@"var jack = {name:'Jack', age:18, car:'BMW'}"];
JSValue *value = context[@"jack"];
JSManagedValue *managedValue = [JSManagedValue managedValueWithValue:value];
context[@"nativeFunction"] = ^(){
   NSLog(@"%@", managedValue.value);
};
```

JSManagedValue 不会对 JSValue 强引用，所以解决了这个循环引用的问题。

5. JSExport

JSExport 可以用来将原生方法输出给 JavaScript，其本身并没有任何属性或方法，需要自己定义一个遵守 JSExport 的协议，然后在协议中声明方法。例如：

```objectivec
// 定义JSDateProtocol暴露给JavaScript可调用的方法
@protocol JSDateProtocol <JSExport>
- (NSString *)getCurrentDate;
@end

// 声明原生类，遵守JSDateProtocol，实现模块的功能
@interface JSDateTool : NSObject <JSDateProtocol>
@end

@implementation JSDateTool
- (NSString *)getCurrentDate {
    NSDate *date = [NSDate date];
    NSDateFormatter *formatter = [[NSDateFormatter alloc] init];
    [formatter setDateFormat:@"yyyy-MM-dd HH:mm:ss"];
    NSString *dateStr = [formatter stringFromDate:date];
    return dateStr;
}
@end

// JavaScript调用Objective-C方法
    JSContext *context = [JSContext new];
    context[@"timeTool"] = [JSDateTool new];
    JSValue *currentTime = [context evaluateScript:@"timeTool.getCurrentDate()"];
    NSLog(@"%@",currentTime); // js call oc
    // 打印:2020-03-07 14:58:46
```

7.2.2 Hermes——Android 的新版 JavaScript 引擎

在 React Native 0.60 版本之前，Android 采用 JavaScriptCore 作为 JavaScript 的引擎。不过由于 JavaScriptCore 是 C++ 编写的，在 iOS 环境下可以直接运行，但在 Android 环境中，则需要 Java 使用关键词 native，它可以用于加载文件和动态链接库，也就是说被 native 修饰的方法可以被 C 语言重写，从而达到与 JavaScriptCore 交互的能力。使用 native 的大致步骤如下。

（1）Java 中声明 native 修饰的方法，类似于声明 abstract 修饰的方法，并没有方法实现，编译该 java 文件，产生一个类文件（.class 文件）。

（2）使用 javah 编译上一步产生的 .class 文件，会产生一个头文件（.h 文件）。

（3）创建一个 .cpp 文件实现 .h 文件中的方法。

（4）将 .cpp 文件编译成动态链接库文件（.dll 文件）。

（5）最后在 Java 中可以使用 System 或是 Runtime 中的 loadLibrary 方法加载 .dll 文件。

例如在 Java 代码中创建了一个类，并定义一个用 native 关键词修饰的方法 printHello，然后在 MainActivity 的 onCreate 方法中调用这个方法并打印到日志中。具体的代码如下。

```java
public class JNIDemo {
    static {
      System.loadLibrary("HelloJNI");
    }

    public static native void sayHello ();

}

public class MainActivity extends AppCompatActivity {
    @Override
    protected void onCreate(@Nullable Bundle savedInstanceState) {
        super.onCreate(savedInstanceState);
        setContentView(R.layout.activity_main);
        String msg = JNIDemo.sayHello();
        Log.e("JNIDemo", msg);
    }
}
```

此时如果直接运行这段代码，系统便会报错，提示找不到"HelloJNI"这个 Library（库）。接下来需要先切换到 JNIDemo.java 所在的目录下，在命令行执行：

```
javac JNIDemo.java
```

然后切换目录到"安卓项目根目录/app/src/main/java"目录下，在命令行执行：

```
javah com.example.JNIDemo
```

执行完毕后可以得到 com_example_JNIDemo.h 头文件，文件的内容如下。

```c
/* DO NOT EDIT THIS FILE - it is machine generated */
#include <jni.h>
/* Header for class com_example_JNIDemo */

#ifndef _Included_com_example_JNIDemo
#define _Included_com_example_JNIDemo
#ifdef __cplusplus
extern "C" {
#endif
/*
 * Class:     com_example_JNIDemo
 * Method:    sayHello
 * Signature: (Ljava/lang/String;)Ljava/lang/String;
 */
JNIEXPORT jstring JNICALL Java_com_example_JNIDemo_sayHello
  (JNIEnv *, jclass);

#ifdef __cplusplus
}
#endif
#endif
```

根据生成的头文件，我们就可以使用C语言实现sayHello方法，回传一个字符串"C: Hello Android!"回去，新建一个HelloJNI.c文件，增加以下代码：

```c
#include <stdlib.h>
#include <stdio.h>
#include "com_example_JNIDemo.h"
#include <string.h>

JNIEXPORT jstring JNICALL Java_com_example_JNIDemo_sayHello
  (JNIEnv *env, jclass cls) {
    char *cstr = "C: Hello Android!";
    return (*env)->NewStringUTF(env,cstr);
}
```

然后我们在main目录下新建一个文件夹jni，将刚刚生成的头文件com_example_JNIDemo.h和HelloJNI.c文件全部复制并粘贴至jni目录下。然后创建一个Android.mk文件，增加以下配置。

```
LOCAL_PATH := $(call my-dir)
include $(CLEAR_VARS)
LOCAL_MODULE    := HelloJNI
LOCAL_SRC_FILES := HelloJNI.c
include $(BUILD_SHARED_LIBRARY)
```

其中LOCAL_MODULE表示要生成的动态链接库so文件的名称，LOCAL_SRC_FILES表示动态链接库中包含的c文件。然后在jni目录下创建Application.mk文件，增加以下配置。

```
APP_ABI := arm64-v8a
APP_PLATFORM := android-16
```

其中APP_ABI表明生成的动态链接库兼容的手机系统架构，APP_PLATFORM表示兼容的Android最低版本。准备过后，我们切换到jni目录下用命令行执行：

```
ndk-build
```

执行过后可以看见在jni目录的同级目录下生成了obj和libs两个文件夹，libs目录下的文件就是编译生成的动态链接库so文件，将so文件放置在app目录下的libs目录中（注意要放置在对应架构的目录下），然后在app目录下的build.gradle文件中增加以下配置。

```
android {

sourceSets {
    main {
        jniLibs.srcDirs = ['libs']
    }
  }

}
```

在gradle.properties文件中增加以下配置。

```
android.useDeprecatedNdk = true
```

至此，运行App并打开MainActivity时，可以看到调用native方法返回的字符串输出到了Android Studio的控制台中。运行的截图如图7.1所示。

```
                2020-03-04 15:39:38.821 26048-26048/? E/JNIDemo: C: Hello Android!
                2020-03-04 15:42:14.410 26048-26048/com.example E/JNIDemo: C: Hello Android!
```

图7.1　Java调用C++模块输出结果

在React Native 0.60之后，React Native推出了自己新的JavaScript引擎——Hermes，针对在Android上运行JavaScript进行了优化，通过对JavaScript在构建时进行Bytecode（字节码）预编译来获取缩短启动时间、减少内存使用量并缩小应用程序大小的效果。具体的使用方式非常简单，只需在Android工程中修改android/app/build.gradle文件。

```
project.ext.react = [
    entryFile: "index.js",
-   enableHermes: false  // clean and rebuild if changing
+   enableHermes: true   // clean and rebuild if changing
]
```

如果之前App已经构建了一次，需要在Android项目根目录下运行./gradlew clean进行清理。

那为什么说Hermes是一个更高效的JavaScript引擎呢？在解释这些问题之前，我们需要先来简单回顾一下JavaScript语言的编译原理。

众所周知，JavaScript是一门解释型语言，与C等典型的变异性语言相比，JavaScript的编译过程通常是在实际执行前进行的，并且不会产生可移植的编译结果。通常来讲，JavaScript的编译过程分为以下几个步骤。

（1）分词与词法分析：把输入的字符串分解为一些对编程语言有意义的代码块（词法单元）。

（2）解析与语法分析：将词法单元集合分析并最终转换为一个由元素逐级嵌套所组成，并且代表了程序语法结构的树，称为AST（Abstract Syntax Tree，抽象语法树）。

（3）代码生成：将AST转换为可执行代码。

整体而言，Hermes并不是一个更"优"的JavaScript引擎，而只是针对移动端场景做了一些关键架构上的取舍，官方也提及并没有计划将该引擎嵌入浏览器或服务器（例如Node.js）架构中。Hermes引擎主要针对内存低、闪存慢等移动端的特点，对以下几个方面进行了改动。

1. 字节码预编译

通常情况下，JavaScript引擎会在代码完全加载完之后再进行语法分析生成AST，然后转化成字节码。而Hermes采取了提前编译的策略，在App构建过程中就开始执行编译命令，为转化字节码提供了更多的时间来进行优化，如图7.2所示。

根据这样的设计，在运行时字节码可以直接映射到内存中且无须读取整个文件，这样避免了在中低端移动端设备性能较差的闪存I/O中整体读取字节码，从而显著提升了App的TTI（Time To Interact，可交互时间）。

2. 移除JIT

JIT（Just-In-Time，即时编译）本身是JavaScript引擎的一个优化策略，它会在引擎中增加一个

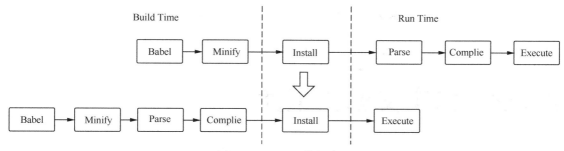

图7.2　JavaScript编译流程

监视器（也叫分析器），监视所有通过解释器解释的代码的运行情况，并记录代码一共运行了多少次、如何运行等信息。如果相同代码运行次数过多，则会标记成"warm"，或逐渐升级成"hot"，之后JIT就把这段代码送到编译器去编译，并且把编译结果存储起来，以便节省之后再进行重复解释的时间。

那Hermes为什么要把JIT移除掉呢？因为JIT虽然做了优化策略，但也不是没有代价，会增加很多其他的开销，例如优化和去优化（将已标记到编译器的代码重新回归到解释器）的开销，监视器记录信息对内存的开销，对基线版本和优化后版本记录的内存开销等。这些在移动端低内存场景下的开销，加上JIT的生效是需要"预热"的（也就是代码逐渐被标记成"warm"或"hot"），这些反而会影响TTI。所以Hermes将自己优化的中心放到了解释器上，而放弃了JIT这种策略。

3. 垃圾回收策略

通常情况下，操作系统的内存管理会采用Swapping的方式优化，将内存中暂时还不能被运行的进程或暂时用不到的程序和数据调到外存上，以便腾出足够的内存供在外存中等待的作业使用。在较为低端的设备中，这样的优化通常并不存在，并且连续的虚拟地址空间通常也都是极为有限的资源，如果使用过多，可能会被操作系统强制杀掉。所以Hermes对内存的垃圾回收策略也做了一定的调整。

（1）按需分配，仅在需要时才以块的形式分配虚拟地址空间。
（2）非连续：虚拟地址空间不限制在单个内存范围内，避免了32位设备上的资源限制。
（3）可移动：可移动意味着可以对内存进行碎片整理，并可将不再使用的块返回给操作系统。
（4）分代：每次垃圾回收时不扫描整个JavaScript堆，减少垃圾回收的时间。

7.3　本章小结

iOS和Android中各自的JavaScript引擎得以使React Native这种跨平台技术方案落地，不同语言之间的通信也是实现UI转化与事件交互的基础。这一座桥梁衔接了JavaScript与iOS/Android，承载了所有消息的通知，所以也就容易成为性能上的瓶颈。熟悉其中的通信模式，有助于我们更好地利用原生能力，也能帮助我们有意识地规避性能上的缺陷。

第8章 自定义原生组件

React Native 自身已经提供了诸多基础组件，例如 View、Text、Image、ScrollView 和 FlatList 等，基本可以满足日常的开发需求。但随着业务的发展，或为了解决性能的瓶颈，我们难免需要自己开发一些原生组件提供给 JavaScript 端使用，例如上文提到的音视频处理和本地存储等。本章将介绍如何使用 React Native 开发定制化的原生 UI 组件，以及如何将组件整理成可方便其他开发者使用的 React Native 插件形式。

8.1 原生 UI 组件封装

React Native 自定义原生组件的方式与上章讲解的自定义模块基本类似，本节将以封装原生的列表组件为例具体说明；鉴于语言和平台本身的差异性，下面将分别描述 iOS 和 Android 是如何实现这一功能的。

8.1.1 iOS 原生组件封装

JavaScript 想要使用原生能力的核心是导出，在这里 React Native iOS 代码提供了一个导出类 RCTViewManager，并且提供了一系列的导出函数用于参数和方法的导出。

首先将需要导出的原生视图对象限定为 UIView 及其子类对象，其次保证每个对象都由一个继承于 RCTViewManager 的 viewManager 对象管理，且在 manager 对象中设置导出模块名和需要导出的方法以及属性，基本用法如下。

```
@interface TableViewManager : RCTViewManager
@end

@implementation TableViewManager
RCT_EXPORT_MODULE(ListView) // 声明导出模块名称，如不声明则为类名
- (UIView *)view {
    return [[TableView alloc] init];
```

```
}
@end

// JavaScript 调用方式
const NativeList = requireNativeComponent("ListView");
......
render() {
    return (
     <NativeList
        ......
     />
    );
}
```

1. 属性传值

上述例子只是简单实现了组件的封装,通常情况下我们需要在 JavaScript 端以 React Props 的形式将数据传递给原生组件。RCT_CUSTOM_VIEW_PROPERTY 是 React Native iOS 提供的用于导出属性给 JavaScript 组件使用的宏,基本用法如下。

```
@implementation TableViewManager
......
RCT_CUSTOM_VIEW_PROPERTY(
   dataSource,  // 属性名称
   NSArray,     // 属性类型
   TableView    // 由 viewManager 导出的 view 实例对象的类
) {
    // 为解决类型不一致的问题这里提供了 RCTConvert 类进行转换,我们可以调用自定义函数修改对应的 UI 展示
    [view reloadDataWithDataSource:[RCTConvert NSArray:json]];
}
......
// JavaScript 调用方式
render() {
  return (
    <NativeList
      style={{ height: DEVICE_HEIGHT-100, width: DEVICE_WIDTH }}
      dataSource={data} // ['第一行', '第二行' ......]
    />
  );
}
```

2. 原生执行JavaScript回调

除了数据之外，React Native Props也有可能传递JavaScript。在原生UI组件接收到用户行为触发的相关事件时，调用该函数处理具体的响应逻辑。

由上述入口文件（TableViewManager）可知，实际的UI部分是交由TableView自定义类负责的，这是一个单纯的继承于UIView的普通视图类，在这个类的代码里我们可以监听列表项的点击事件，然后调用JavaScript函数。

```
#import <React/RCTComponent.h>
@interface TableView : UIView
// React Native提供的block对象，需要注意的是此对象的命名必须以on开头
@property (nonatomic, copy) RCTBubblingEventBlock onClickedItem;
@end

@implementation TableView
……
- (void)tableView:(UITableView *)tableView didSelectRowAtIndexPath:(NSIndexPath *)indexPath {
    // RCTBubblingEventBlock的用法跟block相同，也可以传递参数给JavaScript端（这里用index举例）
    !self.onClickedItem ? : self.onClickedItem(@{@"index": [NSString stringWithFormat:@"%ld", indexPath.row + 1]});
}
……
// 调出onClickedItem方法供JavaScript端使用
@implementation TableViewManager
……
RCT_EXPORT_VIEW_PROPERTY(onClickedItem, RCTBubblingEventBlock)
……

// JavaScript调用方式
render() {
  return (
    <NativeList
      ……
      onClickedItem={({ nativeEvent }) => { console.log(nativeEvent.index) }}
    />
  );
}
```

8.1.2　Android原生组件封装

　　Android端原生UI组件的封装主要分为注册和导出两个部分。一个原生组件必须经过ViewManager的封装导出才能被JavaScript端引入使用。为此，React Native Android代码提供了两种ViewManager抽象类——SimpleViewManager和ViewGroupManager。这两个类均是抽象类，区别是ViewGroupManager相比SimpleViewManager多了提供管理子UI组件的能力。也就是说，如果自定义的原生UI组件经过封装后在JavaScript端还需要嵌套子UI组件，那么就需要继承ViewGroupManager实现抽象方法，否则只需要继承SimpleViewManager即可。SimpleViewManager和ViewGroupManager的继承关系可以由图8.1所示的自定义UI组件相关类中看出来。

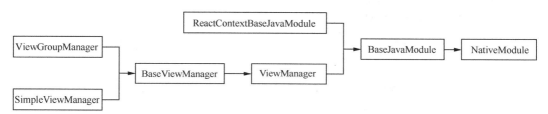

图8.1　React Native Android自定义UI组件相关类

　　SimpleViewManager：接收JavaScript端参数，将原生UI组件与LayoutShadowNode绑定在一起。其中布局相关的参数传递给LayoutShadowNode，LayoutShadowNode则通过开源框架Yoga实现Flex布局。原生UI组件通过这个绑定关系，获得了支持Flex布局的能力。另外，SimpleViewManager会将样式参数及回调事件相关参数传递给原生UI组件，原生UI组件调用原生方法以实现JavaScript端参数配置的效果。

　　ViewGroupManager：除了接收JavaScript端参数和将原生UI组件与LayoutShadowNode绑定之外，还提供了管理子UI组件的能力，使原生UI组件经过封装后能够在JavaScript端支持嵌套子UI组件。

　　BaseViewManager：SimpleViewManager和ViewGroupManager的抽象父类，通过接收JavaScript端参数完成了UI组件部分样式的通用实现，比如设置背景颜色、设置透明度和设置z轴高度等。

　　ViewManager：BaseViewManager的父类，定义了ViewManager类的一些抽象方法，但没有任何实现，具体的实现在子类中。

　　回顾上一章，我们通过继承ReactContextBaseJavaModule实现了自定义原生模块，自定义原生组件通过继承都能被JavaScript端调用，所以它们具有相同的父类BaseJavaModule，且BaseJavaModule实现了NativeModule接口。React Native初始化时，会遍历这些NativeModule类进行实例化。所以不论原生模块还是原生组件，经过封装后都需要在ReactPackage中进行注册，然后再将ReactPackage添加到Application中的ReactNativeHost中。

　　示例代码如下。

```java
// 原生组件封装
public class NativeListViewManager extends SimpleViewManager<RecyclerView> {

    /**
     * @return 原生端导出给 JavaScript 端使用的原生组件名称
     * 可以在JavaScript通过 const NativeList = requireNativeComponent("ListView"); 引
入封装的原生 UI 组件
     */
    @Nonnull
    @Override
    public String getName() {
        return "ListView";
    }

    /**
     * 创建原生 UI 组件实例
     *
     * @param reactContext 上下文对象
     * @return 原生 UI 组件实例
     */
    @Nonnull
    @Override
    protected RecyclerView createViewInstance(@Nonnull ThemedReactContext reactContext) {
        RecyclerView listView = new RecyclerView(reactContext);
        listView.setAdapter(new ListAdapter());
        listView.setLayoutManager(new LinearLayoutManager(reactContext));
        listView.setId(R.id.native_list_view); // 设定 ViewId 用于原生 View 向 React Native 发送事件
        return listView;
    }

    ......

}

// ReactPackage, 可以包含多个原生模块以及自定义UI组件
public class NativeListPackage implements ReactPackage {
    @Nonnull
    @Override
```

```java
        public List<NativeModule> createNativeModules(@Nonnull ReactApplication
Context reactContext) {
            return Collections.EMPTY_LIST;
        }

        @Nonnull
        @Override
        public List<ViewManager> createViewManagers(@Nonnull ReactApplicationContext
reactContext) {
            List<ViewManager> viewManagers = new ArrayList<>();
            viewManagers.add(new NativeListViewManager());
            return viewManagers;
        }
    }

    // Android启动时的Application对象
    public class MainApplication extends Application implements ReactApplication {
        @Override
        public ReactNativeHost getReactNativeHost() {
            return reactNativeHost;
        }

        private final ReactNativeHost reactNativeHost = new ReactNativeHost(this) {
            @Override
            public boolean getUseDeveloperSupport() {
                return BuildConfig.DEBUG;
            }
            // 加入原生ReactPackage，直到这一步才能确保JavaScript端可以调用刚刚声明的原生UI组件
            @Override
            protected List<ReactPackage> getPackages() {
                return Arrays.asList(new NativeListPackage());
            }

            @Override
            protected String getJSMainModuleName() {
                return "index";
            }
        };
    }
```

```
// JavaScript 与 iOS 例子中一致
const NativeList = requireNativeComponent("ListView");
……
render() {
  return (
    <NativeList
      style={{ height: DEVICE_HEIGHT-100, width: DEVICE_WIDTH }}
    />
  );
}
```

1. 属性传值

Android端可以采用@ReactProp注解的方式对方法进行修饰，并且需要设置name。这里我们将name设置为dataSource，表明JavaScript端对应的属性名称为dataSource。值得注意的是，被注解的方法必须具有public void关键词，且方法的入参中第一个参数必须是被封装的原生UI组件实例，后面的参数是想从JavaScript端接收的属性值。这里设置成JavaScript端需要接收一个数组作为数据源，注意不要重复定义相同名称的自定义属性。

```
public class NativeListViewManager extends SimpleViewManager<RecyclerView> {
    ……
    /**
     * 通过 @ReactProp 注解为 JavaScript 端的 UI 组件增加一个自定义参数, name 为参数名称
     * 注解的方法必须具有 public void 关键词
     *
     * @param view 原生 UI 组件实例
     * @param data JavaScript端传入的自定义参数的值
     */
    @ReactProp(name = "dataSource")
    public void setData(RecyclerView view, ReadableArray data) {
        // 将 ReadableArray 对象转化为字符串数组设置给原生列表组件
        ArrayList<String> list = new ArrayList<>();
        for (Object content : data.toArrayList()) {
            list.add((String) content);
        }
        ((ListAdapter) view.getAdapter()).setData(list);
    }
    ……
}
```

```javascript
// JavaScript使用方式
render() {
  return (
    <NativeList
      ......
      dataSource={data} //  ['第一行', '第二行' ...]
    />
  );
}
```

2. 原生执行JavaScript回调

Android端对于JavaScript事件回调的封装分为注册和发送两步，注册时需要指定原生端事件发生时需要发送的原生事件名称，以及JavaScript端接收这个原生事件及数据时需要设定的回调属性名称，例如：

```java
public class NativeListViewManager extends SimpleViewManager<RecyclerView> {
    public static final String EVENT_ON_ITEM_CLICKED = "onClickedItem";
    ......
    /**
     * 注册回调事件
     *
     * @return 原生组件与 JavaScript端的回调事件映射表
     */
    @Override
    @Nullable
    public Map<String, Object> getExportedCustomDirectEventTypeConstants() {
        MapBuilder.Builder<String, Object> builder = MapBuilder.builder();
        // builder 传入的key为原生事件名称，value为JavaScript端映射的回调事件名称
        // 原生事件名称同JavaScript端回调事件名称设置成相同（可不同）
        builder.put(EVENT_ON_ITEM_CLICKED, MapBuilder.of("registrationName",
EVENT_ON_ITEM_CLICKED));
        return builder.build();
    }

    /**
     * 原生接受用户操作事件的类
     */
    class ItemViewHolder extends RecyclerView.ViewHolder {
        private TextView tvContent;
        public ItemViewHolder(@NonNull View itemView) {
```

```java
            super(itemView);
            tvContent = itemView.findViewById(R.id.tv_example_name);
            itemView.setOnClickListener(v -> {   // 列表元素项的点击事件
                // 拼接事件传递的数据
                WritableMap params = Arguments.createMap();
                params.putInt("index", this.getAdapterPosition());
                // 发送点击事件，JavaScript端即可通过回调接收到事件及传递的数据
                mReactContext.getJSModule(RCTEventEmitter.class).receiveEvent(
                    ((View) itemView.getParent()).getId(), // 注意发送事件时使用的
ViewId 必须是正在封装的原生组件的 ViewId，而不是点击到的列表项组件的 ViewId
                    EVENT_ON_ITEM_CLICKED,
                    params);
            });
        }
        /**
         * 绑定数据的方法
         *
         * @param content 绑定的内容
         */
        public void bind(String content) {
            tvContent.setText(content);
        }
    }

    ......
}


// JavaScript调用方式
render() {
  return (
    <NativeList
      ......
      onClickedItem={({ nativeEvent }) => { console.log(nativeEvent .index) }}
    />
  );
}
```

8.1.3　JavaScript直接调用原生组件方法

同上述原生触发JavaScript回调略有不同，这种方式由JavaScript端直接发起调用，而非原生端首

先触发再通知JavaScript端。

1. iOS

iOS的实现方式较为简单,利用React Native iOS提供的宏RCT_EXPORT_METHOD即可。

```objc
@implementation TableViewManager
……
RCT_EXPORT_METHOD(scrollTo:(nonnull NSNumber *)reactTag
                          index:(NSInteger)index)
{
    // 由于继承于RCTViewManager,这里可以直接使用RCTBridge对象,并通过uiManager找到对应的视图
    [self.bridge.uiManager addUIBlock:^(RCTUIManager *uiManager, NSDictionary<NSNumber *,UIView *> *viewRegistry) {
        // 这里可以执行相关的滚动操作
        TableView *tableView = (TableView *)viewRegistry[reactTag];
        [tableView scrollToIndex:index];
    }];
}
……
@end
```

2. Android

Android端对于JavaScript端调用原生UI组件方法的封装也是分为两步——注册和执行。在注册时,需要在getCommandsMap方法中添加JavaScript端组件可以调用的方法名称和指令ID。当JavaScript端调用组件方法时,需要在receiveCommand方法中对JavaScript端传递过来的指令ID进行匹配,若匹配成功,则调用对应原生UI组件的方法。

下例实现了注册滚动方法scrollTo的指令ID,并在JavaScript端调用scrollTo时触发原生列表组件的smoothScrollToPosition方法完成视图滚动的逻辑。

```java
public class NativeListViewManager extends SimpleViewManager<RecyclerView> {
    private static final int COMMAND_ID_SCROLL_TO_INDEX = 666;
    ……
    /**
     * 注册原生 UI 组件可提供的方法映射表
     *
     * @return 方法映射表
     */
    @Override
    public Map<String, Integer> getCommandsMap() {
```

```java
        Map<String, Integer> commandsMap = new HashMap<>();
        // scrollTo 表示 JavaScript 端组件可以调用的方法名称
        // COMMAND_ID_SCROLL_TO_INDEX 表示方法调用时原生端接收到的指令 ID
        commandsMap.put("scrollTo", COMMAND_ID_SCROLL_TO_INDEX);
        return commandsMap;
    }

    /**
     * 原生端对接收到指令ID的处理
     *
     * @param view          原生UI组件
     * @param commandType   JavaScript端调用组件方法时传递给原生端的指令 ID
     * @param args          JavaScript端调用组件方法时,方法的入参
     */
    @Override
    public void receiveCommand(final RecyclerView view, int commandType, @Nullable ReadableArray args) {
        Assertions.assertNotNull(view);
        // 当JavaScript端调用的方法对应的指令 ID 是 COMMAND_ID_SCROLL_TO_INDEX 时
        if (commandType == COMMAND_ID_SCROLL_TO_INDEX) {
            if (args == null) return;
            int position = args.getInt(0);
            // 调用原生列表组件的方法完成滚动逻辑
            view.smoothScrollToPosition(position);
        }
    }
    ……
}
```

3. JavaScript

JavaScript直接调用原生UI方法会略显烦琐,首先要引入两个新的概念——UIManager和findNodeHandle。

UIManager:主要负责管理所有UI相关的类,并且可以提供方法去根据reactTag获得具体组件对象。reactTag由React Native自身负责维护,无须开发者关心。

findNodeHandle:提供根据ref查找对应的reactTag的方法。

具体示例如下。

```
import { UIManager, findNodeHandle } from 'react-native'
......
  onClicked = (nativeData) => {
    // 在这里收到了原生端点击某一行的回调，在nativeEvent中可以解析出原生端传递的数据
    const data = nativeData.nativeEvent ? nativeData.nativeEvent : nativeData;
    console.warn('原生端点击了第' + data['index'] + '行');
    ......
    // 在这里假设需要调用原生的滚动方法，具体如下
    UIManager.dispatchViewManagerCommand(
      findNodeHandle(this.nativeList),
      // ListView为原生UI组件模块名，scrollTo为原生UI组件方法名称
      UIManager.getViewManagerConfig('ListView').Commands.scrollTo,
      // scrollTo方法入参
      [10]
    );
    ......

  render() {
    return (
      <NativeList
        ref={view => this.nativeList = view} // 需要记录引用
        style={{ height: DEVICE_HEIGHT-100, width: DEVICE_WIDTH }}
        dataSource={data}
        onClickedItem={this.onClicked}
      />
    );
  }
```

8.2 自定义插件

  React Native的流行和推广在很大程度上依赖于社区的建设，各式各样的第三方插件填补了很多业务场景中原生能力的缺失。开发React Native插件也存在一定的门槛，需要iOS、Android、JavaScript三端的开发者协同工作，共同制定组件需要实现的功能，定义所需的API及参数。而社区则提供了命令行工具create-react-native-module，来帮开发者生成规范化的第三方插件。

  下面通过npm安装这个命令行工具。

```
npm install -g create-react-native-module
```

然后执行：

```
create-react-native-module MyLibrary
```

即可生成模块名为 MyLibrary 的原生端模块，具体目录结构如下。

```
|--android              // Android模块原生代码
|--ios                  // iOS模块原生代码
|--index.js             // JavaScript端对接代码
|--package.json
```

此时 ios 和 android 目录下的仅仅是原生模块的代码，并不能作为独立的项目运行，所以通常情况下还需要在里面建立一个 React Native 用于开发和调试，例如：

```
|--android              // Android模块原生代码
|--ios                  // iOS模块原生代码
|--index.js             // JavaScript端对接代码
|--example              // 插件示例，也是一个可运行React Native的工程
   |-- android          // Android工程
   |-- ios              // iOS工程
   |-- src              // 插件使用示例代码
   |-- App.js
   |-- index.js
|--package.json
```

8.3 本章小结

React Native 的自定义原生组件机制极大程度上提升了 JavaScript 的能力上限，例如可以通过二次封装提供 React Native 版的地图 SDK，或利用原生绘图的能力提供类似于 SVG 的组件用法。不过这种封装方式并不算是一种高效的手段，需要很大的原生开发成本；并且对于 UI 组件来说，多了一层跨语言交互可能导致一些未知的问题，且开发过程中 JavaScript 开发者与原生开发者之前的协同也是一种额外的成本。

第9章 React Native 的导航方案

在大多数的移动端场景中,我们不可能将所有的业务逻辑都放在一个视图中,用户只要稍加操作就会触发进入下一个视图或回退到上一个视图的动作。在 iOS 和 Android 原生开发中,视图间的前进/后退这一系列的切换操作功能和概念并不统一:iOS 中会使用 UIKit 提供的 UINavigationController 和 UITabBarController 进行视图的导航,并且天然会利用栈的形式存储历史路径;而 Android 本身并不强调路由和导航的概念(虽然会有一些框架去实现这样的功能),页面间的跳转主要通过 new Intent() 的方式实现,通常会传入当前上下文环境和目标环境,但并不关注之前的视图会是什么。

React Native 作为给 Web 开发者使用的工具,除了要兼容上述的两种场景,也需要符合 Web 开发者的开发习惯,避免增加过多的学习成本。在现代前后端分离的 Web 系统中,页面间的跳转通常会交于前端控制,例如在 React 或 Vue 的开发生态中,都会提供对应的路由模块进行页面级组件的切换,也就是 URL 的跳转。毕竟对于 Web 系统来说,只要用户在浏览器里访问了这个 URL,都应该获取到这个 URL 对应的页面,而不用去理会用户之前访问的历史路径是什么。

React Native 自身并没有提供导航的解决方案,通常我们会采用社区中的两种主流方案:一是使用 JavaScript 实现的 react-navigation,二是利用原生实现的 react-native-navigation。本章将概述这两种方案实现的原理和适用的场景,以及在更为烦琐的混合应用中如何去应对原生和 React Native 的路由。

9.1 原生导航偏好

在不同平台上,各系统对导航的实现也有偏好,例如 iOS 通常会在底部使用 UITabBarController,然后每个 Tab 会对应独立的 UINavigationController(导航栏控制器)作为一个视图管理,并使用栈的方式管理普通的 UIViewController,如图 9.1 所示。

Android 则倾向于使用侧边的 DrawerLayout,这是 Google 推出的侧边菜单的控件,侧边菜单可以根据手势展示与隐藏,当手指从屏幕左侧向右滑动时会出现一个抽屉式的菜单从左侧展示出来,并且是以类似于悬浮的效果展示在主界面上,充分利用了设备上有限的空间,如图 9.2 所示。

图9.1　iOS TabBar 导航　　　　　　　图9.2　Android DrawerLayout 导航

9.2　JavaScript导航——React Navigation

React Navigation是一个核心功能均由JavaScript实现的导航模块，包含了页面间的切换、动画和缓存等一系列功能。对于原生应用来说，这一用法仅仅是渲染了一个原生视图ViewController（iOS）/Activity（Android），后续的导航变化均在此视图中实现，原生端并不能、也不用监听其变化，只需要提供相关的手势拦截（react-native-gesture-handler），避免触发原生的交互即可。具体视图的结构和关系如图9.3所示。

图9.3　React Navigation视图结构关系

9.2.1 自定义导航

React Navigation中主要包含了Stack、Tab和Drawer等多种常用的导航方式（4.x以上的版本将这几种方式抽取成了独立的模块，如需使用则要单独安装），大致的使用方式如下。

```
"react-native-gesture-handler": "^1.5.6",
"react-native-reanimated": "^1.7.0",
"react-native-safe-area-context": "^0.7.2",
"react-native-screens": "^2.0.0-beta.2",
"@react-native-community/masked-view": "^0.1.6",
"@react-navigation/bottom-tabs": "^5.0.2",
"@react-navigation/native": "^5.0.1",
"@react-navigation/stack": "^5.0.1",

// 最常见的Stack导航方式
// Home.js 视图组件
import React from 'react';
import { View, Text } from 'react-native';

export default class Home extends React.Component {
  render() {
    return (
      <View style={{ flex: 1, alignItems: 'center', justifyContent: 'center' }}>
        <Text>Home Screen</Text>
      </View>
    );
  }
}


// App.js
import React from 'react';
import { NavigationContainer } from '@react-navigation/native';
import { createStackNavigator } from '@react-navigation/stack';
import Detail from './Page/Detail'
import Home from './Page/Home'

const HomeStack = createStackNavigator();
function App() {
  return (
    <NavigationContainer>
```

```
      <HomeStack.Navigator>
        <HomeStack.Screen name="Home" component={Home} />
        <HomeStack.Screen name="Detail" component={Detail} />
      </HomeStack.Navigator>
    </NavigationContainer>
  );
}
export default App;

// index.js
import { AppRegistry } from 'react-native';
import App from './App';

AppRegistry.registerComponent('ReactNavigation', () => App);
```

通过这样声明式的方式就可以获得一个Stack导航，且每个被配置的视图组件会被注入navigation对象，用于进行页面间的跳转和获取参数等行为，例如：

```
class Home extends React.Component {
  render() {
    const { navigation } = this.props;
    return (
      <View style={{ flex: 1, alignItems: 'center', justifyContent: 'center' }}>
        <Text>{ navigation.state.params.id }</Text>
        <Button
          title="跳转"
          onPress={() => {
            this.props.navigation.navigate('Detail', {
              id: 1,
            });
          }}
        />
      </View>
    );
  }
}
```

React Navigation并没有使用原生导航的API，所以用户可以任意定制导航页面和规则，但随之加重的是JavaScript的线程压力，这也是在策略上做出的取舍。

除此之外，Stack 导航通常也会和 Tab 和 Drawer 导航结合使用，以实现 iOS 和 Android 上最常见的布局方式，例如：

```
// 底部Tab导航
import React from 'react';
import { NavigationContainer } from '@react-navigation/native';
import { createStackNavigator } from '@react-navigation/stack';
import { createBottomTabNavigator } from '@react-navigation/bottom-tabs';

......
// 省略了视图组件的声明

const Tab = createBottomTabNavigator();

const HomeStack = createStackNavigator();
const MineStack = createStackNavigator();

const HomeScreen = () => (
  <HomeStack.Navigator>
    <HomeStack.Screen name="Home" component={Home} />
    <HomeStack.Screen name="Detail" component={Detail} />
  </HomeStack.Navigator>
)

const MineScreen = () => (
  <MineStack.Navigator>
    <MineStack.Screen name="Mine" component={Mine} />
    <MineStack.Screen name="Setting" component={Setting} />
  </MineStack.Navigator>
)

function App() {
  return (
    <NavigationContainer>
      <Tab.Navigator>
        <Tab.Screen name="HomeTab" component={HomeScreen} />
        <Tab.Screen name="MineTab" component={MineScreen} />
      </Tab.Navigator>
    </NavigationContainer>
  );
```

```
}

export default App;
```

实际效果如图9.4所示。

图9.4 React Navigation TabBar示例

```
// 左侧抽屉式导航
import React from 'react';
import { Text, View } from 'react-native';
import { createAppContainer } from 'react-navigation';
import { createDrawerNavigator } from 'react-navigation-drawer';

……
class Home extends React.Component {
  static navigationOptions = {
    drawerLabel: 'Home',
  };

  render() {
    return (
      <View style={{ flex: 1, alignItems: 'center', justifyContent: 'center' }}>
        <Text>Home</Text>
      </View>
```

```
    );
  }
}
……

const HomeStack = createStackNavigator({
  Home: Home,
  Detail: Detail,
});

const AccountStack = createStackNavigator({
  Setting: Setting,
  Info: Info,
});

const DrawerNavigator = createDrawerNavigator({
  Home: HomeStack,
  Account: AccountStack,
});

export default createAppContainer(TabNavigator);
```

实际效果如图9.5所示。

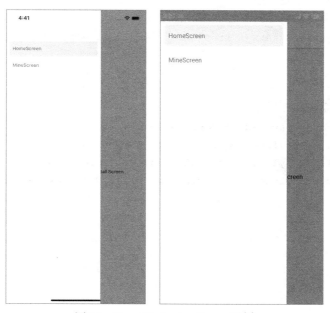

图9.5　React Navigation Drawer示例

9.2.2 导航事件

通过React自身的生命周期，我们可以获取视图组件的创建和销毁的时机，以便进行业务的处理。但移动端所需的页面生命周期通常会超出React本身的范畴，例如页面显示viewDidAppear（iOS）/onResume（Android）和页面隐藏viewWillDisappear（iOS）/onPause（Android）。React Navigation通过事件的方式补充了这一缺失的功能，可以在组件内手动注册事件，也可以利用高阶组件<NavigationEvents />或React Hooks——useFocusState进行状态的监听。具体的事件类型分为以下4种：willFocus/didFocus/willBlur/didBlur，其中did前缀的事件主要表示如果页面跳转中存在过渡动画的话，确认动画结束后才会触发didFocus/didBlur，示例如下。

```
// 手动注册事件
const didFocusListener = this.props.navigation.addListener(
  'didFocus',
  payload => {
    console.log('didFocus', payload);
  }
);
……
didFocusListener.remove()

// 使用高价组件
import React from 'react';
import { View } from 'react-native';
import { NavigationEvents } from 'react-navigation';

const Screen = () => (
  <View>
    <NavigationEvents
      onWillFocus={payload => console.log('will focus', payload)}
      onDidFocus={payload => console.log('did focus', payload)}
      onWillBlur={payload => console.log('will blur', payload)}
      onDidBlur={payload => console.log('did blur', payload)}
    />
    ……
  </View>
);

// 使用React Hooks
function Screen() {
  const focusState = useFocusState();
```

```
return (
  <View>
    <Text>{focusState.isFocused}</Text>
    <Text>{focusState.isBlurring}</Text>
    <Text>{focusState.isBlurred}</Text>
    <Text>{focusState.isFocusing}</Text>
  </View>
)
}
```

9.3 原生导航——React Native Navigation

React Native Navigation是社区中另一款主流的导航解决方案，采用了与React Navigation完全不同的方式，页面间的切换均由原生端实现，与常规的原生App在视图关系上基本一致。

iOS视图关系如图9.6所示。

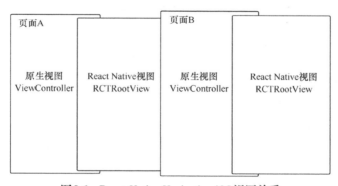

图9.6　React Native Navigation iOS视图关系

Android视图关系图9.7所示。

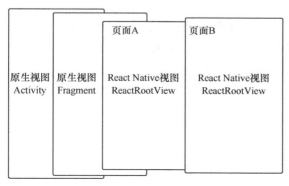

图9.7　React Native Navigation Android视图关系

这样对于整个项目而言，视图的结构和机制与原生端差别不大，只不过 React Native Navigation 提供了一个虚拟"ViewController"的概念，自己实现了一套视图挂载、销毁和切换的效果，抹去了 iOS 和 Android 上的差异。不过由于视图的控制权仍在原生端，因此 React Native Navigation 依然可以监听原生端特有的生命周期及其他事件（例如屏幕旋转等），并且利用事件通信同步给 JavaScript 端。

9.3.1 自定义导航

在 React Native Navigation 中，React 组件需要先注册成视图，然后使用配置的方式定义具体的导航关系，例如：

```
// index.js
import { Navigation } from 'react-native-navigation';
import HomeScreen from './Home.js'

Navigation.registerComponent('HomeScreen', () => HomeScreen);

Navigation.events().registerAppLaunchedListener(() => {
  Navigation.setRoot({
    root: {
      stack: {
        children: [{
          component: {
            name: 'HomeScreen'
          }
        }]
      }
    }
  });
});
```

其中 Navigation.events().registerAppLaunchedListener 在 App 启动时注册了监听函数，并通过执行 Navigation.setRoot 创建了视图布局。除了 Stack 之外，React Native Navigation 也提供了其他常见的布局，例如底部 Tab 导航和侧边栏，具体用法如下：

```
// 底部 Tab 导航
Navigation.setRoot({
  root: {
    bottomTabs: {
      children: [
        {
```

```
            component: {
              name: 'Home',
            },
          },
          {
            component: {
              name: 'Mine',
            },
          },
        ],
      },
    }
});

// 侧边栏
Navigation.setRoot({
  root: {
    sideMenu: {
      left: {
        component: {
          name: 'Menu',
        }
      },
      center: {
        stack: {
          children: [{
            component: {
              name: 'ChatList',
            },
          }],
        }
      },
    }
  }
});
```

实际效果如图9.8所示。

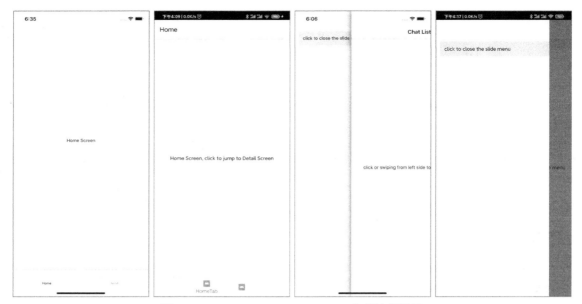

图9.8　React Native Navigation TabBar与侧边栏效果

通常情况下，我们会将一种或一种以上的导航方式组合起来使用，例如：

```
// 组合布局
Navigation.setRoot({
  root: {
    bottomTabs: {
      children: [{
        stack: {
          children: [
            {
              component: {
                name: 'HomeScreen',
              }
            },
            {
              component: {
                name: 'DetailScreen',
              }
            }
          ],
          options: {
            bottomTab: {
```

```
            text: 'Home',
          }
        }
      }
    },
    {
      component: {
        name: 'Mine',
        options: {
          bottomTab: {
            text: 'Mine',
          }
        }
      }
    }]
  }
});
```

另外，Navigation提供了push、pop等API来处理页面间的跳转逻辑，例如：

```
// Home.js

Navigation.push(this.props.componentId, {
  component: {
    name: 'DetailScreen',
    passProps: {
      text: 'Message from Home'
    },
  }
});

// Detail.js
……
// 在Detail页面可以通过Props获取到Home页面传入的参数
this.props.text
……

// pop
Navigation.pop(this.props.componentId);
```

9.3.2 视图生命周期

由于React Native Navigation的导航和视图机制完全依赖于原生，因此原生视图触发相关生命周期时只需要通知React对应的视图组件即可，具体代码如下。

```
class Home extends Component {

  componentDidMount() {
    this.navigationEventListener = Navigation.events().bindComponent(this);
  }

  componentWillUnmount() {
    if (this.navigationEventListener) {
      this.navigationEventListener.remove();
    }
  }

  componentDidAppear() {
    console.log('Home did appear');
  }

  componentDidDisappear() {
    console.log('Home did disappear');
  }
}
```

除了单视图的监听之外，Navigation.events()还提供了全局监听的方式，例如：

```
const didAppearListener = Navigation.events().registerComponentDidAppearListener(
  ({ componentId, componentName, passProps }) => {
    ……
  }
);

const DidDisappearListener = Navigation.events().registerComponentDidDisappearListener(
  ({ componentId, componentName }) => {
    ……
  }
);
```

9.4 混合导航探索

在实际业务场景和历史项目中,我们可能会遇到更为烦琐的场景:或是原生App中需要部分使用React Native提升开发效率,或是React Native项目中需要唤起原生视图(例如地图)提供原生能力支持。这样就会在导航栈中形成React Native视图和原生视图混排的情况,处理起来会相当烦琐,特别是还存在视图跳转时传递参数的场景。上文提到的两种导航方案都是从完全React Native的项目的角度进行设计,对原生和React Native混合的场景并没有做太多的考虑。如果出现以下的几种情况,现有的方案可能无法直接支持。

(1)在React Native项目中,调起一个原生视图B后再调起一个React Native视图C,页面C越过原生视图B,直接回退到初始的React Native视图A中,具体结构如图9.9所示。

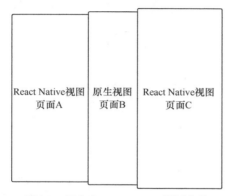

图9.9　原生和React Native视图交替

(2)原生App视图页面A中有一部分页面B是由React Native渲染的,以该视图B作为入口,后续调起的一系列React Native视图均可直接回退到页面A,具体结构如图9.10所示。

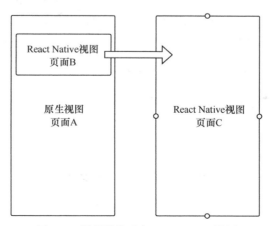

图9.10　局部渲染唤起React Native视图

对于这类React Native和原生视图混排的场景，我们该如何调整导航方案呢？

9.4.1 方案设计

由于React Native绘制的视图的本质是原生视图，且原生端对于所有视图的进出和跳转有着绝对的控制，能通过非侵入业务的方式监听整个App视图的变化，而JavaScript端则无法感知这一切，因此我们将导航的历史记录存储在原生一侧，只有当跳转到React Native视图时才将信息同步给JavaScript端。具体流程如图9.11所示。

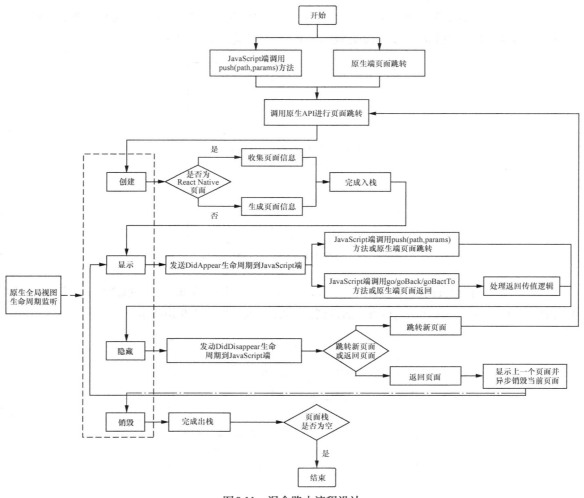

图9.11 混合路由流程设计

1. iOS

iOS路由流程设计如图9.12所示。

第9章 React Native的导航方案

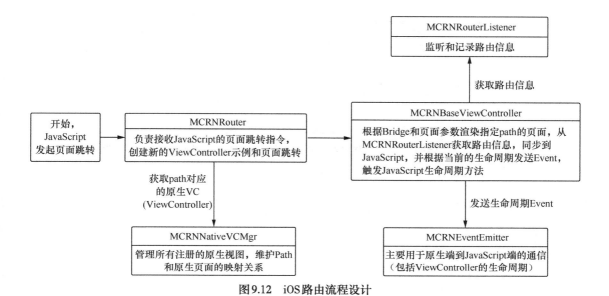

图9.12 iOS路由流程设计

下面以其中的MCRNBaseViewController和MCRNRouter的关键代码为例进行说明。

```
// MCRNBaseViewController

@interface MCRNBaseViewController ()<UINavigationControllerDelegate, UIGesture
RecognizerDelegate>
......
@end


@implementation MCRNBaseViewController

// 初始化视图，记录当前ViewController对应的bridge、路径名称和参数，以及唯一标识viewId
- (instancetype)initWithBridge:(RCTBridge *)bridge componentName:(NSString *)
componentName path:(nullable NSString *)path query:(nullable id)query viewId:
(NSString *)viewId {
    self = [super init];
    if (self) {
        self.bridge = bridge;
        self.componentName = componentName;
        self.path = path;

        if ([query isKindOfClass:[NSDictionary class]]) {
            self.query = [(NSDictionary *)query jsonString];
        } else if ([query isKindOfClass:[NSString class]]) {
```

```objc
            self.query = query;
        }
        // 视图唯一标识
        if (viewId) {
            self.viewId = viewId;
        } else {
            NSString *uuid = [[NSUUID UUID] UUIDString];
            self.viewId = [NSString stringWithFormat:@"MCNative.%@", uuid];
        }
    }
    return self;
}

- (void)viewDidLoad {
    [super viewDidLoad];
    ……
    // 初始化RCTRootView
    self.rootView = [[RCTRootView alloc] initWithBridge:self.bridge moduleName:self.componentName initialProperties:paramM.copy];
    [self.view addSubview:_rootView]; // 将RCTRootView挂载到ViewController中
    ……
}
……
@end


// MCRNRouter
@implementation MCRNRouter

RCT_EXPORT_MODULE()

// JavaScript端发出的Push操作
RCT_EXPORT_METHOD(push:(NSString *)componentName path:(NSString *)path query:(NSString *)query viewId:(NSString *)viewId nativeConfig:(NSString *)config) {
    if ([componentName isEqualToString:@"Native"]) {
        // 根据参数可以选择跳转到纯粹原生页面
        ……
    } else {
        // 跳转到由React Native 绘制视图的ViewController
        NSDictionary *configDict = [NSString dictionaryWithJsonString:config];
```

```objc
        RCTBridge *currentBridge = self.bridge;
        if ([currentBridge isKindOfClass:NSClassFromString(@"RCTCxxBridge")]) {
            currentBridge = [self.bridge valueForKey:@"parentBridge"];
        }
        // 初始化MCRNBaseViewController
        MCRNBaseViewController *rnVC = [[MCRNBaseViewController alloc] initWithBridge:currentBridge componentName:componentName path:path query:query viewId:viewId];

        rnVC.nativeConfig = configDict;
        rnVC.hidesBottomBarWhenPushed = true;
        BOOL animated = true;
        if (configDict[@"animation"] != nil) {
            animated = [configDict[@"animation"] boolValue];
        }
        // Push新页面入栈
        [[self navigationViewController] pushViewController:rnVC animated:animated];
    }
}

// JavaScript端发出的goBack操作
RCT_EXPORT_METHOD(goBack:(NSString *)data) {
    // 获取当前路由栈中的ViewController
    NSArray<UIViewController *> *vcStack = [self navigationViewController].viewControllers;
    if ([vcStack.lastObject isKindOfClass:[MCRNBaseViewController class]] && ((MCRNBaseViewController *)vcStack.lastObject).isPresent) {
        [self pop]; // pop当前ViewController
        [[self navigationViewController] popViewControllerAnimated:false];
    } else {
        [[self navigationViewController] popViewControllerAnimated:true];
    }
    [self sendDataToTargetVC:[self navigationViewController].viewControllers.lastObject data:data];
}

@end
```

2. Android

Android路由流程设计如图9.13所示。

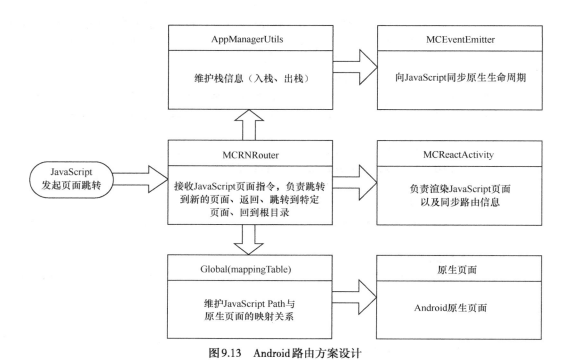

图9.13 Android路由方案设计

同样,下面也对其中用于处理视图渲染和路由管理的核心类 MCReactActivity 和 MCRNRouter 进行具体的说明。

```java
// MCReactActivity.java
public class MCReactActivity extends BaseActivity
        implements DefaultHardwareBackBtnHandler, PermissionAwareActivity,
IMCCurrentRouter {
    ......
    private MCReactActivityDelegate createMCReactActivityDelegate() {
        return new MCReactActivityDelegate(this, getCurrentRouter().getComponentName()) {
            @Override
            protected Bundle getLaunchOptions() {//同步路由信息
                Bundle bundle = new Bundle();
                bundle.putString("componentName", getCurrentRouter().getComponentName());
                ......
                bundle.putString("routesInfo", AppManagerUtils.getInstance().getRoutesInfo().toString());
```

```java
            return bundle;
        }
    };
}

@Override
protected void onCreate(Bundle savedInstanceState) {
    if (mDelegate == null) {
        if (savedInstanceState == null) { // 初次打开页面,从Intent中获取页面信息
            routerBean.setComponentName(getIntent().getStringExtra("componentName"));
            ......
        } else { // Activity 可能被内存回收后再次重建,此时从销毁前暂存的数据恢复
            routerBean = (MCRouterBean) savedInstanceState.getSerializable(MC_SAVE_INSTANCE_BEAN);
        }
        ......
        super.onCreate(savedInstanceState);
        // 创建代理,MCReactActivityDelegate继承于ReactActivityDelegate
        // ReactActivityDelegate实现了具体的页面渲染等逻辑
        mDelegate = createMCReactActivityDelegate();
    }
    mDelegate.onCreate(savedInstanceState);
}
......
}


//MCRNRouter.java
public class MCRNRouter implements IMCRouter {
......
public void push(String componentName, String path, String query, String viewId, String nativeConfig) {
        ......
        // 根据全局的路由管理类获取当前Activity信息
        Activity activity = AppManagerUtils.getInstance().mStackBean.lastElement().getmActivity();
        if (componentName.equals("Native")) { // 根据参数可以选择跳转到纯粹原生页面

            ......
            Intent intent = new Intent();
```

```
            ......
            activity.startActivity(intent);
        } else {
            try {
                ......
                Intent intent = new Intent();
                // 跳转一个新的MCReactActivity
                intent.setClass(activity, MCReactActivity.class);
                ...... // 传参
                activity.startActivity(intent);
                ......
            } catch (Exception e) {
                Log.e(TAG, "push: error----->" + e.toString());
            }
        }
    }
    ......
}
```

3. JavaScript

JavaScript 的主要作用是指定路由信息，映射路由与组件的关系，再根据原生同步的路由信息渲染对应的视图，大致的关键流程如下。

```
const Router = ({ routes }) => {
  history.initRoutesConfig(routes);   // 初始化所有的路由信息
  return (WrappedComponent) => (
    class WrapperComponent extends React.Component {
      render() {
        const { viewId, componentName, routesInfo, path, query } = this.props;
        // 获取原生端传入的路由信息，进行同步
        history.syncRoutes(toJSON(routesInfo), componentName);
        return (
          <RouterContext.Provider
            value={{ history, routes, viewId, path, query }}
          >
            <WrappedComponent { ...this.props } />
          </RouterContext.Provider>
        )
      }
```

```js
    });
};

export default class Route extends React.PureComponent {
  render() {
    const { children } = this.props;
    return (
      <RouterContext.Consumer>
      {
        ({ history, routes, viewId }) => {
          const route = history.getRoute(viewId);  // 获取视图配置信息
          ......
          let View = routes[route.path];  // 获取当前视图组件
          ......
          return (
            <View
              ......
            >
              { children }
            </View>
          )
        }
      }
      </RouterContext.Consumer>
    )
  }
}

// 根视图
class App extends Component {
  render() {
    return (
      <Route />
    );
  }
}
// 注册路由信息
AppRegistry.registerComponent('mcrn', () => Router({
  routes: {
    '/main': Main,   // 页面组件
```

```
    '/page': Page,    // 页面组件
  },
})(App));
```

整体案例及使用方法可参考电子资源。

9.4.2 扩展功能

除了共享导航信息之外，我们还可以利用这种机制将原生视图的特性同步给JavaScript。例如在JavaScript中直接调起原生视图，监听原生视图生命周期，监听手势、物理按键或屏幕旋转等无法被JavaScript感知的原生事件。

1. 调起原生视图

在React Native中调起原生视图并不是一个很复杂的事，只是根据不同的业务场景需要做额外的定制开发，且一旦场景数量增多，势必会增加代码的维护成本，所以我们将需要被调起的原生视图统一注册到导航模块。具体的实现类似于React Native的自定义模块，通过iOS的宏以及Android添加视图类的方式，在导航模块中标记有哪些路径对应的是原生视图，这样在JavaScript中就可以通过唯一路径调起原生视图，具体步骤如下。

在iOS端注册一个可被唤起的原生页面。

```
#import <MCRNBridge/MCRNBridge.h>

@implementation YourViewController
MCRNRegisterNative('NativeView')   // 注册一个路由名称为NativeView的原生视图
... YourCode

- (void)dealloc {
  MCRNNativeDealloc
}
@end
```

在Android端注册一个可被唤起的原生页面。

```
// ... MainApplication.java
private Map<String, Class> getNativeClass() {
    Map<String, Class> map = new HashMap<>();
    map.put("'NativeView'", FirstActivity.class);
    return map;
}
```

```java
@Override
public void onCreate() {
    super.onCreate();
    SoLoader.init(this, /* native exopackage */ false);
    MCRNBridgeManager
            .getInstance()
            .initApplication(this, getNativeClass())  / 将原生页面名称映射到React Native中
            .create("MCBundle", getReactNativeHost())
            .activate("MCBundle");
}
```

在JavaScript端的调用方式为：

```
import { history } from '@mcrn/bridge'

history.push('NativeView', { msg: 'msgFromMain' }, 'Native')
```

2. 扩展原生视图属性

同React Native Navigation一样，扩展JavaScript视图原生能力都是通过在React Native视图组件中生成唯一标志并创建事件监听，以及在原生视图触发生命周期或其他事件时通过EventEmitter，同步给对应的React Native视图实现的，例如：

```objc
// MCRNBaseViewController 监听页面的原生生命周期
@implementation MCRNBaseViewController

- (void)viewDidAppear:(BOOL)animated {
    [super viewDidAppear:animated];
    // ...
    // 发送viewDidAppear生命周期到JavaScript端
    [MCRNLifeCycleEventEmitter sendLifeCycleWithType:COMPONENT_DID_APPEAR to:self.viewId data:nil];

}

// MCRNLifeCycleEventEmitter 原生广播，负责将生命周期通知给与JavaScript通信的模块
@implementation MCRNLifeCycleEventEmitter
// 发送页面生命周期
+ (void)sendLifeCycleWithType:(LifeCycleType)type to:(NSString *)viewId data:(nullable NSString *)data{
    ……
```

```objc
    // 发送原生通知把消息转到MCRNEventEmitter
    [[NSNotificationCenter defaultCenter] postNotificationName:kMCRNEventInternalCommunicationKey object:msgDict.copy];
}
@end

// MCRNEventEmitter 与JavaScript通信的模块，接收原生中广播的事件，传递给JavaScript端
@implementation MCRNEventEmitter

RCT_EXPORT_MODULE()

- (instancetype)init {
    self = [super init];
    if (self) {
        // 注册原生通知
        [[NSNotificationCenter defaultCenter] addObserver:self selector:@selector(didReceiveNoti:) name:kMCRNEventInternalCommunicationKey object:nil];
    }
    return self;
}
// 发送来自原生的通知到JavaScript端
- (void)didReceiveNoti:(NSNotification *)noti {
    if ([noti.object isKindOfClass:[NSDictionary class]]) {
        NSDictionary *msgDict = (NSDictionary *)noti.object;
        // Event发送代码
        [self sendEventWithName:msgDict[@"eventName"] body:msgDict[@"body"]];
    }
}

@end


// Android 端关键代码

// 当原生页面状态为不可见时
@Override
public void onHostPause() {
    ......
    StackBean routerBean = AppManagerUtils.getInstance().findViewIdFromActivity(getCurrentActivity());
```

```java
        if (routerBean == null) return;
        MCEventEmitter.getInstance().emitComponentDidDisappear(routerBean.getViewId(),
routerBean.getComponentName());
    }

    // MCEventEmitter.java
    // 通知JavaScript 原生Activity 执行OnPause
    @Override
    public void emitComponentDidDisappear(String viewId, String componentName) {
        WritableMap event = Arguments.createMap();
        event.putString("componentName", componentName);
        event.putString("viewId", viewId);   // 视图唯一标识
        MCEventEmitterUtils.getInstance().emit(ComponentDidDisappear, event,
componentName);
    }

    public void emit(String eventName, WritableMap data, String componentName) {
        emitToJs(eventName, data, componentName);
    }

     private void emitToJs(String eventName, Object data, String componentName) {
     ......
        reactContext
         .getJSModule(DeviceEventManagerModule.RCTDeviceEventEmitter.class)
         .emit(eventName, data);
    }
```

另外,对于Android独有的生命周期,例如onActivityResult(用于接收上个页面回退到当前页面时的数据),也可以在iOS上做到兼容,模拟出一样的效果,例如:

```objc
//  MCRNRouter
@implementation MCRNRouter
// JavaScript端发出的回退操作
RCT_EXPORT_METHOD(goBack:(NSString *)data) {
//  ......
    // 发送需要回传到上个页面的数据
    [self sendDataToTargetVC:targetVC data:data];
}

- (void)sendDataToTargetVC:(UIViewController *)target data:(NSString *)data {
```

```
        ......
        // 将数据使用 MCRNLifeCycleEventEmitter 以Event的形式发送给JavaScript端
        if ([target isKindOfClass:[MCRNBaseViewController class]]) {
            [MCRNLifeCycleEventEmitter sendLifeCycleWithType:COMPONENT_RECEIVE_
RESULT to:((MCRNBaseViewController *)target).viewId data:data];
        }
        ......
    }
    @end
```

9.5 本章小结

除了评估适用场景之外，React Native项目导航方案的选择在一定程度上也需要考虑自身的开发能力。如果具备一定的原生能力，选取以原生导航为基础的方案可以更好地利用原生特性，并且可以很顺畅地衔接现有的原生应用，以及在其基础上做扩展。如果是纯粹的JavaScript开发者，可以选取完全使用JavaScript开发的导航方案，控制所有相关的生命周期及事件。

第10章　热更新与多实例

React Native项目打包编译后会生成一个或多个JavaScript文件和静态资源，将其添加到原生工程，再由React Native原生部分通过对应的项目路径对其进行加载、解析，然后渲染成具体的视图，这就是目前最常见的部署过程。在这个过程中，如果生成的JavaScript文件和静态资源不是一开始就放置到原生工程，而是通过网络请求下载到本地目录，那React Native是否还能进行正常的解析？另外，如果一个工程中存在多个React Native的Bundle.js文件，能否按照一定的业务逻辑去加载不同的文件，从而呈现不同的场景？带着这几个问题，本章将介绍在React Native中如何实现热更新，介绍对React Native App平台化的探索——多个React Native实例的管理。

10.1　热更新

热更新是App开发中永远绕不开的一个话题，特别是在混合开发的模式下，大家都希望能快速响应需求，减少App从开发完成到上架的时间。不管是最初的WebView还是如今的React Native，都希望通过网络的能力下载自己的业务代码，并在App中直接生效。前面已经讲解了React Native的启动过程，原生端生成具体的React Native实例（RCTBridge/ReactNativeHost）需要加载被指定路径的JavaScript文件，通过更换该路径下的JavaScript文件并生成新的React Native实例就可以实现热更新的效果。

10.1.1　热更新流程

在讨论具体的实现之前，先梳理一下热更新的具体流程，如图10.1所示。

通常进行热更新检测的时机在React Native启动之后（除非项目首页是原生页面，在React Native初始化之前先做了热更新检测），也就是说此时React Native实例已经生成。如果JavaScript Bundle文件由于更新而产生变化，则App需要重新加载React Native实例并重新渲染视图。React Native提供了重启自身实例的API，它可以通过此API或通过重启App的方式更新React Native实例及之前已经渲染成功的页面，且能够尽可能避免之前实例生成的旧数据和旧视图的遗留而可能产生的问题。

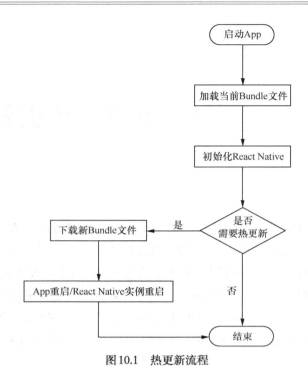

图10.1　热更新流程

10.1.2　第三方服务

热更新的整体方案需要包含服务端和原生端的逻辑，从检查更新到下载文件（可能还包含解压文件），都不是能单纯通过JavaScript解决的。目前市面上也有可以直接使用的热更新服务，例如微软的CodePush，它相当于一个中心仓库，开发者可以上传需要更新的文件（包括JavaScript、HTML、CSS及图片等文件），然后在App中检测是否存在文件更新，如果存在的话则将最新的文件下载到本地。在使用CodePush服务之前，需要安装它的命令行工具NodeJS-based CLI（可以通过npm install -g code-push-cli命令进行安装），然后注册CodePush账号及你的App即可。对于React Native，CodePush也提供了对应的插件react-native-code-push，这样安装成功后就可以直接使用热更新服务了，例如：

```
import React, { Component } from 'react';
import {
  Platform,
  StyleSheet,
  Text,
  View
} from 'react-native';
import CodePush from "react-native-code-push";
```

```javascript
const codePushOptions = {
  checkFrequency : CodePush.CheckFrequency.ON_APP_RESUME    // 检查更新的频率
};

class App extends Component {

  componentDidMount() {
    CodePush.sync({
      // 对话框
      updateDialog: {
        appendReleaseDescription: true,
        descriptionPrefix: '更新内容：',
        title: '更新',
        mandatoryUpdateMessage: '',
        mandatoryContinueButtonLabel: '更新',
      },
      mandatoryInstallMode: CodePush.InstallMode.IMMEDIATE, // 安装模式
      deploymentKey: '',   // iOS Key，部署环境(Production/Staging)
    });
  }

  render() {
    return (
      ……
    );
  }
}

App = CodePush(codePushOptions)(App)

export default App
```

10.1.3 具体实现

如果需要自行实现热更新服务，除了在服务端做好版本控制外，在原生端也有一些关键流程需要注意，本小节就分以下3个方面阐述热更新的具体实现。

1. 配置Bundle文件路径

React Native 默认会在特定路径加载特定名字的 JavaScript Bundle 文件，例如 iOS 的 Main Bundle 和 Android 的 asset 路径下的 index.android.bundle 文件；同时也支持自定义 Bundle 路径，在实例化时可以指

定目录加载自定义名称的Bundle文件,例如:

```objc
// iOS自定义JavaScript Bundle路径
@interface Foo : NSObject <RCTBridgeDelegate> // 遵守RCTBridgeDelegate协议
@end


@implementation Foo

// 实现RCTBridgeDelegate协议方法
- (NSURL *)sourceURLForBridge:(RCTBridge *)bridge
{
  NSURL *bundleURL = ...; // 这里是自定义的JavaScript Bundle URL
  return bundleURL;
}


@end
```

```java
// Android 自定义JavaScript Bundle路径
private final ReactNativeHost mReactNativeHost = new ReactNativeHost(this) {
    ......
    @Nullable
    @Override
    protected String getJSBundleFile() { // 重写该方法来改变JavaScript Bundle加载路径
        String jsBundleFile = getFilesDir().getAbsolutePath() + "/index.android.bundle";
        File file = new File(jsBundleFile);
        return file.exists() ? jsBundleFile : null;
    }
    ......
};
```

2. 下载文件

React Native本身对文件处理没有提供太多的支持,如果想要实现检测下载新的JavaScript Bundle文件,就必须使用第三方插件进行支持,或者直接在原生端开发下载文件功能。

iOS中最常见的网络工具就是AFNetworking了,下面简单展示一下其基本的用法。

```objc
@implementation MCAmpBundleHelper
```

```objc
- (void)downloadBundleForURL:(NSURL *)URL
                         MD5:(NSString *)MD5
                  resHandler:(BundleDownloadBlock)resHandler {
    
    // 这里使用了iOS最常见的网络工具 AFNetworking
    NSURLSessionConfiguration *configuration = [NSURLSessionConfiguration defaultSessionConfiguration];
    AFURLSessionManager *manager = [[AFURLSessionManager alloc] initWithSessionConfiguration:configuration];
    NSURLRequest *request = [NSURLRequest requestWithURL:URL];
    
    NSURLSessionDownloadTask *downloadTask = [manager downloadTaskWithRequest:request progress:^(NSProgress * _Nonnull downloadProgress) {
        NSLog(@"下载进度%@", downloadProgress);
    } destination:^NSURL *(NSURL *targetPath, NSURLResponse *response) {
        // 返回下载目标路径，这里放在沙盒中
        return [NSURL fileURLWithPath:[NSTemporaryDirectory() stringByAppendingPathComponent:[NSString stringWithFormat:@"%@", [response suggestedFilename]]]];
    } completionHandler:^(NSURLResponse *response, NSURL *filePath, NSError *error) {
        if (error) {
            // 下载过程抛出异常
            resHandler(nil, error);
        } else {
            // JavaScript Bundle文件下载成功
            resHandler(filePath, nil);
        }
    }];
    // 发起下载
    [downloadTask resume];
}

@end
```

在Android中，实现文件下载的方式比较多，可以采用网络请求框架如OKHTTP、Volley，或者直接使用HttpClient、HttpURLConnection等方式，这里简单介绍一下使用HttpURLConnection下载文件的关键方法。

```java
// downloadUrl 为文件的URL地址
// saveFile 为Android文件对象
public boolean downloadUpdateFile(String downloadUrl, File saveFile) throws Exception {
```

```java
        long totalSize = 0;
        int updateTotalSize = 0;
        HttpURLConnection httpConnection = null;
        InputStream is = null;
        FileOutputStream fos = null;
        try {
            URL url = new URL(downloadUrl);
            httpConnection = (HttpURLConnection) url.openConnection();
            httpConnection.setRequestProperty("Accept-Encoding", "identity");
            httpConnection.setConnectTimeout(10000);
            httpConnection.setReadTimeout(20000);
            httpConnection.connect();
            updateTotalSize = httpConnection.getContentLength();
            if (httpConnection.getResponseCode() == 404) {
                throw new Exception("fail!");
            }
            is = httpConnection.getInputStream();
            fos = new FileOutputStream(saveFile, false);
            byte buffer[] = new byte[1024];
            int readsize = 0;
            while ((readsize = is.read(buffer)) != -1) {
                fos.write(buffer, 0, readsize);    // 写入文件内容
                totalSize += readsize;
            }
        } catch (Exception e) {// 下载失败
            totalSize = 0;
        } finally {
            if (httpConnection != null) {
                httpConnection.disconnect();
            }
            if (is != null) {
                is.close();
            }
            if (fos != null) {
                fos.close();
            }
        }
        return totalSize != 0 && (totalSize == updateTotalSize);
    }
```

3. 重启实例

下载JavaScript Bundle完成之后，就是重启React Native实例了。在iOS中，我们可以通过直接调用React Native Bridge提供的方法重启这个实例。

```
// 获取需要重启的Bridge实例
RCTBridge *bridge = ...;
// 重新加载Bundle 完成重启
[bridge reload];
```

Android虽然也提供了重启方法，例如：

```
// Android React Native 实例重启代码
// ReactNativeHost 中
ReactInstanceManager reactInstanceManager= MainApplication.getInstance().getReactNativeHost().getReactInstanceManager();
reactInstanceManager.recreateReactContextInBackground();
```

但实际效果只是重新创建了React Native的上下文，之前依赖于旧实例的视图需要开发者自行手动销毁，否则会产生未知的错误。所以最简单的方式是重启应用，重新走一遍加载和渲染的流程，确保不会存在脏数据。

10.2 App平台化——React Native多实例

在如今一些体量大、业务线多的场景中，App在一定程度上承载了更多入口的作用，看上去更像一个平台而非单一的业务应用。例如美团、淘宝这类App，打开后可以看到里面包含了多个不同种类的业务，各自使用的频次不尽相同，高频的如点外卖，低频的如订机票等。这样对于用户而言，下载一个App就可以满足大部分的日常需求，不需要为了每年一两次的使用再去下载其他的App，统一的账号体系省去了注册等重复的流程。对于开发者而言，不同的业务通常由不同的产研团队维护，这些业务如何在同一个App中共同运作，互不干扰？这就给App的架构设计带来新的问题和挑战。

对于这种平台化的需求，原生App有自己的解决之道，iOS通过pod引入，Android可以拆成多个颗粒度更细的apk，这些都可以很好地达到预期的效果，那React Native中是否也有类似的解决方案呢？

10.2.1 多实例管理

对于React Native项目来说，平台化方案最基础的诉求就是需要在同一个App中同时共存多个React Native实例，也就是存在多个Bundle.js文件。对于这种情况，我们必然需要一个管理类对这些实例进行统一的控制，设计方案可以参考图10.2。

```
┌─────────────────────────────────┐
│    ReactNativeBridgeManager     │
├─────────────────────────────────┤
│ + bridgeConfigMap               │
│ + birdgeMap                     │
│ ......                          │
├─────────────────────────────────┤
│ + createBridgeForConfig         │
│ + registerBridge                │
│ + bridgeForBundleName           │
│ ......                          │
└─────────────────────────────────┘
```

图 10.2　多实例管理类

bridgeConfigMap：主要包含了创建实例所需的配置信息，并与 BundleName 对应。

bridgeMap：BundleName 与 Bridge 的映射集合。

createBridgeForConfig：通过配置信息创建实例。

registerBridge：直接注册已经生成的 Bridge 对象。

bridgerForBundleName：根据 BundleName 获取对应的 Bridge。

10.2.2　指定渲染依赖实例

每个 React Native 视图的渲染都依赖于当前的 React Native 实例，也就是通过 JavaScript Bundle 文件生成的对象，其中 iOS 中命名为 bridge，而 Android 中则称之为 ReactNativeHost，例如：

```
// iOS RCTRootView.m
......
// RCTRootView初始化
- (instancetype)initWithBridge:(RCTBridge *)bridge   // bridge即为React Native实例
                    moduleName:(NSString *)moduleName
              initialProperties:(NSDictionary *)initialProperties
{
  RCTAssertMainQueue();
  RCTAssert(bridge, @"A bridge instance is required to create an RCTRootView");
  RCTAssert(moduleName, @"A moduleName is required to create an RCTRootView");

  RCT_PROFILE_BEGIN_EVENT(RCTProfileTagAlways, @"-[RCTRootView init]", nil);
  if (!bridge.isLoading) {
    [bridge.performanceLogger markStartForTag:RCTPLTTI];
  }

  if (self = [super initWithFrame:CGRectZero]) {
    self.backgroundColor = [UIColor whiteColor];
  // 在当前视图类中保存对应的React Native实例
    _bridge = bridge;
    _moduleName = moduleName;
```

```objc
    _appProperties = [initialProperties copy];
    _loadingViewFadeDelay = 0.25;
    _loadingViewFadeDuration = 0.25;
    _sizeFlexibility = RCTRootViewSizeFlexibilityNone;

    // 注册相关的事件通知
    [[NSNotificationCenter defaultCenter] addObserver:self
        selector:@selector(bridgeDidReload)
        name:RCTJavaScriptWillStartLoadingNotification
                                                object:_bridge];

    [[NSNotificationCenter defaultCenter] addObserver:self
        selector:@selector(javaScriptDidLoad:)
        name:RCTJavaScriptDidLoadNotification
                                                object:_bridge];

    [[NSNotificationCenter defaultCenter] addObserver:self
        selector:@selector(hideLoadingView)
        name:RCTContentDidAppearNotification
                                                object:self];

#if TARGET_OS_TV
    self.tvRemoteHandler = [RCTTVRemoteHandler new];
    for (NSString *key in [self.tvRemoteHandler.tvRemoteGestureRecognizers allKeys]) {
        [self addGestureRecognizer:self.tvRemoteHandler.tvRemoteGestureRecognizers[key]];
    }
#endif

    [self showLoadingView];

    // Immediately schedule the application to be started.
    // (Sometimes actual '_bridge' is already batched bridge here.)
    // 调用bundleFinishedLoading，传入React Native实例（RCTBridge）来渲染具体的UI视图（RCTContentView）
    [self bundleFinishedLoading:([_bridge batchedBridge] ?: _bridge)];
    }

    RCT_PROFILE_END_EVENT(RCTProfileTagAlways, @"");
```

```objc
    return self;
}

……

- (void)bundleFinishedLoading:(RCTBridge *)bridge
{
  RCTAssert(bridge != nil, @"Bridge cannot be nil");
  if (!bridge.valid) {
    return;
  }

  [_contentView removeFromSuperview];
  // _contentView即为通过React Native实例渲染的实际UI视图
  _contentView = [[RCTRootContentView alloc] initWithFrame:self.bounds
                                                    bridge:bridge
                                                  reactTag:self.reactTag
                                            sizeFlexiblity:_sizeFlexibility];
  [self runApplication:bridge];

  _contentView.passThroughTouches = _passThroughTouches;
  [self insertSubview:_contentView atIndex:0];

  if (_sizeFlexibility == RCTRootViewSizeFlexibilityNone) {
    self.intrinsicContentSize = self.bounds.size;
  }
}
```

```java
// Android ReactActivityDelegate.java
public class ReactActivityDelegate {
  ……
  // 获取在Application中初始化的ReactNativeHost
  protected ReactNativeHost getReactNativeHost() {
    return ((ReactApplication) getPlainActivity().getApplication()).getReactNativeHost();
  }
  ……
  protected void loadApp(String appKey) {
    if (mReactRootView != null) {
```

```java
            throw new IllegalStateException("Cannot loadApp while app is already running.");
        }
        // 通过ReactNativeHost渲染具体的UI视图
        mReactRootView = createRootView();
        mReactRootView.startReactApplication(
            getReactNativeHost().getReactInstanceManager(),
            appKey,
            getLaunchOptions());
        getPlainActivity().setContentView(mReactRootView);
    }
}
```

可以看出,两端代码在渲染UI时依赖的React Native实例和App初始化时的实例基本进行了绑定,并没有预留可动态变化的空间。因此,在平台化方案中,我们需要修改甚至重写部分代码,将渲染所依赖的实例修改为上述管理类中的方法,这样确保开发者在渲染React Native视图中可以指定使用哪一个实例,而不是固定成唯一一个,具体参考代码如下。

1. iOS

重写基础视图类 ViewController,关键代码如下。

```objc
// MCRNBaseViewController.m
// 初始化增加bundleName参数,可指定多实例中具体启用的实例名称
- (instancetype)initWithBundleName:(NSString *)bundleName
                     componentName:(NSString *)componentName
                              path:(nullable NSString *)path
                             query:(nullable id)query
                            viewId:(nullable NSString *)viewId;
{
    RCTBridge *bridge = [[MCRNBridgeMgr shared] bridgeForBundleName:bundleName];
    NSAssert(bridge, @"bundleName: %@ 未在MCRNBridgeMgr中注册", bundleName);
    return [self initWithBridge:bridge componentName:componentName path:path query:query viewId:viewId];
}
```

2. Android

重写基础视图类的代理类 ReactActivityDelegate.java,关键代码如下。

```java
// MCReactActivityDelegate.java
public abstract class MCReactActivityDelegate extends ReactActivityDelegate {
    ......
```

```
        // 覆盖ReactActivityDelegate中的getReactNativeHost写法,将其交由MCRNBridgeManager
控制
        @Override
        public ReactNativeHost getReactNativeHost() {
            return MCRNBridgeManager.getInstance().get(mBundleName);
        }
    }
```

10.2.3 自定义原生模块依赖

除了渲染视图外,和React Native实例有依赖关系的还包括自定义原生模块和原生UI组件。这两种原生能力的添加方式前面已经讲解过,那在这种多实例的场景下原有的添加方式是否会改变呢?

1. iOS

在前面的章节中讲解了iOS端如果需要添加原生依赖模块就要用到一个宏——RCT_EXPORT_MODULE,导出当前的类供JavaScript端调用,那在多实例场景下还有什么其他的影响吗?下面看一下这个宏大致的实现方式。

在RCTBridgeModule.h中RCT_EXPORT_MODULE宏是这样定义的:

```
#define RCT_EXPORT_MODULE(js_name) \
RCT_EXTERN void RCTRegisterModule(Class); \
+ (NSString *)moduleName { return @#js_name; } \
+ (void)load { RCTRegisterModule(self); }
```

可以看到这个宏实现了两个类方法+(NSString *)moduleName和+(void)load,前者负责返回当前模块的名称,后者则调用了RCTRegisterModule,其作用及关键代码如下。

```
static NSMutableArray<Class> *RCTModuleClasses;

// 获取所有注册的原生模块
NSArray<Class> *RCTGetModuleClasses(void)
{
    __block NSArray<Class> *result;
    dispatch_sync(RCTModuleClassesSyncQueue, ^{
        result = [RCTModuleClasses copy];
    });
    return result;
}
```

```objc
// 注册原生模块
void RCTRegisterModule(Class moduleClass)
{
  static dispatch_once_t onceToken;
  dispatch_once(&onceToken, ^{  // 这个代码块中的代码只会执行一次
    RCTModuleClasses = [NSMutableArray new];
    RCTModuleClassesSyncQueue = dispatch_queue_create("com.facebook.react.ModuleClassesSyncQueue", DISPATCH_QUEUE_CONCURRENT);
  });

  RCTAssert([moduleClass conformsToProtocol:@protocol(RCTBridgeModule)],
        @"%@ does not conform to the RCTBridgeModule protocol",
        moduleClass);

  // Register module,
  dispatch_barrier_async(RCTModuleClassesSyncQueue, ^{
   // 将类(原生模块)添加到RCTModuleClasses数组中
    [RCTModuleClasses addObject:moduleClass];
  });
}
```

可以看出，React Native维护了RCTModuleClasses的静态变量，且RCTModuleClasses只会被实例化一次，并且提供了注册和获取原生模块的方法。

最后看一下RCTCxxBridge.mm文件中的代码片段。

```objc
@implementation RCTCxxBridge

- (void)start
{
  // 调用RCTGetModuleClasses, 获取所有原生模块进行挂载
  (void)[self initializeModules:RCTGetModuleClasses() withDispatchGroup:prepareBridge lazilyDiscovered:NO];
  // ……
}
@end
```

每个Bridge实例都会在初始化方法中调用start方法，由此可知，当拥有多个RCTBridge实例时，所有实例都会挂载上RCT_EXPORT_MODULE宏导出的原生模块。

2. Android

在Android中，用户需要手动将自定义模块添加到ReactNativeHost中，所以分属不同Bundle的自定义原生模块可以非常简单地就归属于不同的ReactNativeHost，且互不影响。例如：

```java
// 初始化ReactNativeHost
private final ReactNativeHost mReactNativeHost = new ReactNativeHost(this) {
    @Override
    public boolean getUseDeveloperSupport() {
        return BuildConfig.DEBUG;
    }

    @Override
    protected List<ReactPackage> getPackages() {
        return Arrays.<ReactPackage>asList(
            new MainReactPackage(),
            // 添加所需的自定义模块
            ……
        );
    }

    @Override
    protected String getJSMainModuleName() {
        return "index";
    }
};

@Override
public ReactNativeHost getReactNativeHost() {
    return mReactNativeHost;
}

……
// 初始化另一个ReactNativeHost
private final ReactNativeHost otherReactNativeHost = new ReactNativeHost(this) {
    @Override
    public boolean getUseDeveloperSupport() {
        return false;
    }

    @Override
```

```java
    protected List<ReactPackage> getPackages() {
        return Arrays.<ReactPackage>asList(
            new MainReactPackage(),
            // 添加所需的自定义模块
            ……
        );
    }

    @Nullable
    @Override
    protected String getBundleAssetName() {
        // 配置不同的Bundle文件名
        return "index.android.bundle";
    }
};

@Override
public ReactNativeHost getOtherReactNativeHost() {
    return otherReactNativeHost;
}
……

@Override
public void onCreate() {
    super.onCreate();
    SoLoader.init(this, /* native exopackage */ false);
    // 使用管理类托管两个ReactNativeHost
    MCRNBridgeManager
            .getInstance()
            .create("bundle", getReactNativeHost())
            .create("otherBundle", getOtherReactNativeHost())
            .activate("bundle"); // 激活指定的ReactNativeHost
}
```

10.2.4 多实例效果及局限

实现了上述功能后，基本上就可以在原生环境下使用两个不同的React Native实例了。我们可以在页面跳转时多加一个参数指定对应Bundle名称，例如：

```
//JavaScript React Native 多实例跳转用法
```

```
import { history } from '@mcrn/bridge'

history.push('/about', {msg: 'msgFromMain'}, '', { navBarTextColor: 'black'},
'bundle2')
```

1. iOS

```
// MCRNRouter.m
......
RCT_EXPORT_METHOD(push:(NSString *)componentName path:(NSString *)path query:
(NSString *)query viewId:(NSString *)viewId nativeConfig:(NSString *)config
bundleName:(NSString *)bundleName) {
......
// 根据传入的BundleName获取对应的bridge, 并进行渲染
MCRNBaseViewController *rnVC;
if (bundleName && [[MCRNBridgeMgr shared] bridgeForBundleName:bundleName]) {
    rnVC = [[MCRNBaseViewController alloc] initWithBridge:[[MCRNBridgeMgr shared]
bridgeForBundleName:bundleName] componentName:componentName path:path query:query
viewId:viewId];
} else {
    // 当前bridge
    RCTBridge *currentBridge = self.bridge;
        if ([currentBridge isKindOfClass:NSClassFromString(@"RCTCxxBridge")]) {
            currentBridge = [self.bridge valueForKey:@"parentBridge"];
        }
    rnVC = [[MCRNBaseViewController alloc]
}
......
}
```

2. Android

```
// MCRNRouter.java
......
public void push(String componentName, String path, String query, String viewId,
String nativeConfig, String bundleName) {
    ......
    Intent intent = new Intent();
    intent.setClass(activity, MCReactActivity.class);
    // 多加一个BundleName即可, MCReactActivityDelegate会根据其中的BundleName匹配到对
应的ReactNativeHost进行视图渲染
```

```
intent.putExtra(bundleName, bundleName);
……
activity.startActivity(intent);
}
```

不过目前这种方案有一定的局限性，例如两个Bundle依赖的React Native版本必须一致，单个Bundle增加原生依赖必须告之原生App并手动添加，暂时无法做到完全的自动化及流程化。

10.3 本章小结

无论是热更新还是多实例，本质上都是在React Native实例的使用方式上做了扩展。只有熟悉、理解了React Native运行的机制及原理，再结合原生的能力，开发者才能在架构使用层面对React Native使用的场景进行进一步的探索和扩展。

第11章 常见场景优化

随着版本的更新，React Native 的性能也在一步步提升，但对于一个影响范围甚广的框架而言，保持整体逻辑的一致性、通用性会被放在更高的优先级。所以对于一些特定场景，通常需要开发者自己去实现进一步的优化，这也要求开发者对框架本身的实现原理有一定的了解。本章会分析几个常见的优化场景，并借着这些场景对 React Native 本身的实现机制做进一步的解释。

11.1 页面启动白屏时间

如何降低页面启动的白屏时间，是每个前 Web 开发者都要考虑的问题，大家都希望用户能够更快地看到产品界面并进行接下来的交互。等待的时间越长，用户就越有可能流失。为了缩短首页的白屏时间，第一步就是确认哪些因素会影响页面渲染的时间。React Native 的启动过程是，加载 JavaScript 文件，启动 React Native 实例，最后生成原生视图并挂载到系统屏幕。在这个过程中，有哪些环节会随着业务复杂度的提升而发生变化的呢？本节将对其进行具体的分析并得出量化结果。

11.1.1 JavaScript Bundle 包大小的影响

在 React Native Release 模式下加载 JavaScript Bundle 文件，并使用多个不同大小的 Bundle 文件，观测其对应用冷启动时间的影响。

JavaScript 代码打包 Release 模式命令如下。

```
// iOS
react-native bundle \
    --reset-cache \
    --platform ios \
    --dev false \
    --entry-file index.js \
    --bundle-output ./build/ios/index.ios.jsbundle \
    --assets-dest ./build/ios \
    --sourcemap-output ./build/map/index.ios.jsbundle.map
```

```
// Android
node node_modules/react-native/local-cli/cli.js bundle \
    --reset-cache \
    --platform android \
    --dev false \
    --entry-file index.js \
    --bundle-output ./android/app/src/main/assets/index.android.bundle \
    --assets-dest ./android/app/src/main/res/ \
    --sourcemap-output ./android/app/src/main/assets/index.android.bundle.map
```

表11.1主要记录了JavaScript Bundle文件大小不同时,在iOS和Android上启动React Native所用时长。

表11.1

JavaScript Bundle 文件大小(MB)	启动耗时(ms)(iOS)	启动耗时(ms)(Android)
1.6	557	1067
2.4	669	1256
3.2	781	1483
4.8	926	1851
7.2	1270	2428

从结果来看,JavaScript Bundle文件大小的增加对启动耗时的影响基本呈线性关系,且幅度高于启动耗时。

11.1.2 自定义原生模块的影响

目前React Native并没有开放动态加载模块的方法,在初始化实例时会将所有的自定义原生模块全部加载到当前实例,这样是否会带来一些性能上的消耗,原生模块的数量又是否会对React Native的初始化时间造成影响?下面将对以下几种情况进行测试以说明这几个问题。

1. iOS

前面的章节讲解了RCT_EXPORT_MODULE可以在iOS端导出原生的模块,并且是在RCTBridge初始化时。表11.2记录了不同数量(只列举了一些数量)原生模块加载所需的时间。

表11.2

原生模块数量	启动耗时(ms)
4	30.05
28	28.12
68	30.29
89	31.2
185	33.52

2. Android

Android端在React Native实例启动时会遍历原生模块列表，将自带原生模块及自定义原生模块进行初始化并注册JavaScript端可以调用的常量和方法。为了测试自定义原生模块数量对React Native App启动时间的影响，表11.3记录了启动不同数量（只列举了一些数量）原生模块的情况下，React Native App的启动耗时如下。

表11.3

原生模块数量	启动耗时（ms）
23	1044.5
33	1078.1
43	1079.0
63	1154.8
103	1240.3

可以发现，虽然耗时随着原生模块数量的增加而增加，但在数量级上并没有明显的变化，也就说明这一段流程并没有太多可优化的空间。

11.1.3 页面层级深度

这里的页面层级深度主要指的是基础视图UI嵌套的层级，Web开发者可以理解成DOM的嵌套层级。一般来讲，层级嵌套越深，绘制引擎的计算量就越大，消耗的性能和花费的时间也会增多。React Native在绘制图层时主要依赖Facebook自研的Yoga框架，将React Native的视图转化成对应的原生视图，但在iOS和Android上这套机制并不是完全一致的，所以下面将分开进行分析。

1. iOS

在iOS上，JavaScript中的嵌套布局会如实反馈在原生视图中。例如，当一个View组件嵌套了Text组件，且View组件只是起布局作用时，利用布局分析器可以得到一个图11.1所示的视图层级。

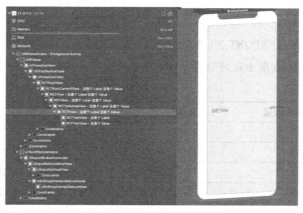

图11.1　iOS布局分析器效果

2. Android

Yoga在Android上除了处理React Native视图到原生UI组件的转化，还对Flex布局嵌套做了优化处理，例如：

```
<View
    style={{
        flexDirection: 'row',
        paddingHorizontal: 15,
        width: '100%',
        justifyContent: 'space-between',
        alignItems: 'center'
    }}>
    <Text>这是个 Label</Text>
    <Text>这是个 Value</Text>
</View>
```

在上面的示例代码中，两个Text组件外面嵌套了一层View组件，组件为它们设置了横向排布、水平边距等一系列布局参数。运行Android的布局分析器可以看到图11.2所示的快照截图。

图11.2　Android布局分析器效果

可以看到,右侧的屏幕中只有两个Text组件,这两个Text组件的UI布局符合我们对JavaScript端的预期。左侧的原生UI组件树形图也表明本页面(ReactRootView)中只有两个ReactTextView。那么前面明明在JavaScript端嵌套了一层View组件作为两个Text组件的父组件,为什么在Android端就没有了呢?

查阅Android端的React Native源码,可以发现在 com.facebook.react.uimanager.NativeViewHierachyOptimizer.java 中定义了如何将只有布局参数的冗余嵌套组件去除的逻辑,核心代码如下。

```java
@Override
public NativeKind getNativeKind() {
    // 如果是虚拟节点或本节点只具有布局相关属性,就返回 NONE;否则返回本节点应该是原生视图树结
    构中的叶子节点还是父节点
    return isVirtual() || isLayoutOnly()
        ? NativeKind.NONE
        : hoistNativeChildren() ? NativeKind.LEAF : NativeKind.PARENT;
}

public void handleCreateView(
    ReactShadowNode node,
    ThemedReactContext themedContext,
    @Nullable ReactStylesDiffMap initialProps) {
    ......

    boolean isLayoutOnly =
        node.getViewClass().equals(ViewProps.VIEW_CLASS_NAME)
            && isLayoutOnlyAndCollapsable(initialProps); // 检查是否只有布局相关的属性
    node.setIsLayoutOnly(isLayoutOnly);

    if (node.getNativeKind() != NativeKind.NONE) {
        // 如果是虚拟节点或本节点只具有布局相关属性,则不会创建对应的原生视图
        mUIViewOperationQueue.enqueueCreateView(
            themedContext, node.getReactTag(), node.getViewClass(), initialProps);
    }
}

private void applyLayoutBase(ReactShadowNode node) {
    ......

    while (parent != null && parent.getNativeKind() != NativeKind.PARENT) {
        if (!parent.isVirtual()) {
```

```
        // 遍历节点时，如果本节点的父节点只具有布局相关属性，那就把父节点的布局属性直接赋值到
子节点上
        x += Math.round(parent.getLayoutX());
        y += Math.round(parent.getLayoutY());
    }
    parent = parent.getParent();
}
applyLayoutRecursive(node, x, y);
```

上述流程也可以这样理解：由JavaScript端定义的布局结构，通过Yoga框架在Android端生成一个对应的ReactShadowNode树形结构，Android端再将这些ReactShadowNode结构进行遍历，如果某些ReactShadowNode只具有布局相关参数，没有任何的UI属性，那么它的布局参数将被分别赋值给它的子节点，然后将它跳过，不为它生成对应的原生UI组件。优化流程的示意图如图11.3所示。

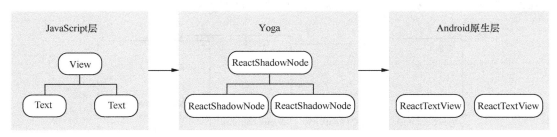

图11.3 Yoga在Android上的优化流程

可以发现，在页面中通过嵌套和平铺两种方式放置同样数量的组件时，加载页面的时间差距不大，所以影响页面加载的主要因素是视图的数量。因为平铺布局时的冗余嵌套开发者很容易发现，但在嵌套布局的情况下，不容易发现这些冗余的组件，所以开发者还是应该尽量避免这样的情况。如果在开发中仅仅想通过嵌套实现某种布局，例如实现Text组件在布局中垂直居中，那么记得不要在这种嵌套的View组件上添加边框、阴影、圆角等非布局属性，否则布局将不会被Yoga框架自动优化，且影响页面渲染时间。

11.2 视图预加载

按需加载和预加载经常用到的两个优化性能方式：前者根据当前的需要加载资源，将暂时用不到的资源延后加载；后者则预判用户下一步的操作，将尚未使用的资源提前加载，以便用户在进入下一步场景时有更好的体验。乍听之下这话略有点矛盾，但联系实际场景就能很好地理解它们的含义。

按需加载通常应对的场景是项目启动后的首次加载，避免加载过多首页、首屏上用户看不到、使用不上的功能，以最小的资源代价最快呈现可操作内容。

预加载的使用场景是，当前页面已满足用户响应及需求，需要进一步预判及优化，但原则上提前加载不能消耗过多的流量（避免产生过多资费）及计算资源（导致当前页的卡顿），不然适得其反。

落实到React Native项目中，按需加载对应的是缩短首屏启动时间，主要依靠减小JavaScript包的大小以及减少自定义模块的数量。而预加载则可以运用到使用原生路由，或存在从原生页面跳转到React Native页面的项目中，提前加载下一个场景的页面，以达到快速响应用户的目的。下面将具体分析其实现的方式。

1. 实现思路

在了解原生导航下React Native视图的层级关系的基础上，开发者只需要做到在A页面时提前创建B页面，但不真正挂载到屏幕上，当用户真正触发操作进入B页面时，才将已经创建好的页面挂载到屏幕上。这也就节省了创建页面这一阶段的时间，避免出现过长时间的白屏。

在iOS中需要提前创建的是RCTRootView对象，如图11.4所示。

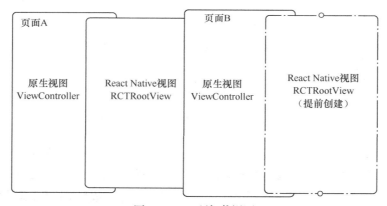

图11.4　iOS预加载视图

在Android中需要提前创建的则是ReactRootView对象，如图11.5所示。

需要注意的是，这种思路并不适用于使用React Navigation作为路由方案的项目。React Navigation自始至终只存在一个RCTRootView/ReactRootView对象，它的路由完全依赖于JavaScript自主控制View组件的切换。所以，如果想在React Navigation中实现预加载，可以考虑在JavaScript端利用transform方式在屏幕可见区域外提前创建视图，从而达到预期的效果。

2. iOS

由于iOS自身会对未展示到屏幕上的内容进行优化，也就是并不真正执行渲染以避免资源的浪费，因此只提前创建RCTRootView并不能达成预先渲染的效果，还需要将RCTRootView实例添加到主窗口上。下面是预加载页面的核心代码。

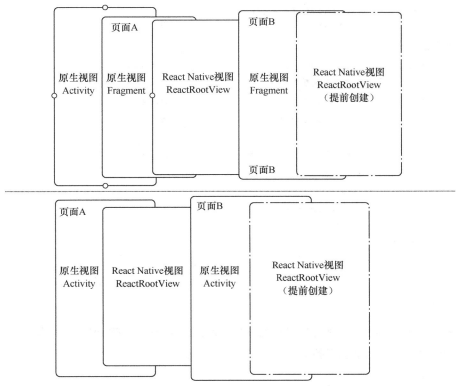

图 11.5 Android 预加载视图

```
@implementation MCRNPreloadViewMgr
// 预加载方法
- (void)preloadWithBridge:(RCTBridge *)bridge componentName:(NSString *)
componentName path:(NSString *)path query: (NSString *)query viewId:(NSString *)
viewId recycle:(BOOL)recycle {

    // ...
    // 提前创建 RCTRootView
    RCTRootView *rootView = [[RCTRootView alloc] initWithBridge:bridge moduleName:
componentName initialProperties:paramM.copy];

    // 添加到窗口的最底层，达到当前页面不可见的效果
    [[UIApplication sharedApplication].keyWindow.rootViewController.view
insertSubview:rootView atIndex:0];
    rootView.frame = [UIScreen mainScreen].bounds;
    // 加入缓存
    [self cacheRootViewWithModal:[[MCRNPreloadModel alloc] initWithPath:path
```

```objc
viewId:viewId rootView:rootView recycle:recycle]];
    }
    // 对外暴露地获取RCTRootView API
    - (RCTRootView *)rootViewForViewId:(NSString *)viewId {
            // 根据viewID查找匹配的RCTRootView
        for (MCRNPreloadModel *model in self.preloadViewCache) {
            if ([model.viewId isEqualToString:viewId]) {
                if (!model.recycle) {
                    // 对于不重复使用的预加载视图，可以将其从缓存池中移除
                    [self.preloadViewCache removeObject: model];
                }
                return model.rootView;
            }
        }
        return nil;
    };
@end
```

当真正执行新页面跳转时，优先从缓存池中查找是否有匹配的RCTRootView，如果有，便直接将已经渲染好的RCTRootView添加到新的MCRNBaseViewController上。

```objc
@implementation MCRNRouter
// ...
// JavaScript端发出的Push操作
RCT_EXPORT_METHOD(push:(NSString *)componentName path:(NSString *)path query:(NSString *)query viewId:(NSString *)viewId nativeConfig:(NSString *)config) {

            // 从缓存池中查找有没有匹配的预加载RCTRootView
            RCTRootView *rootView = [[MCRNPreloadViewMgr shared] rootViewForPath: path] ?: [[MCRNPreloadViewMgr shared] rootViewForViewId:viewId];
            if (rootView) {
                // 将预加载的视图传入新的VC，并添加到View上
                MCRNBaseViewController *rnVC = [[MCRNBaseViewController alloc]
    initPreloadWithBridge:self.bridge componentName:componentName path:path query:query viewId:viewId rootView: rootView];
                [[self navigationViewController] pushViewController:rnVC animated: animated];
                return;
            }
            // ...
```

```objc
        }
    }

    @end


    @implementation MCRNBaseViewController

    - (instancetype)initPreloadWithBridge:(RCTBridge *)bridge componentName:
(NSString *)componentName path:(nullable NSString *)path query:(nullable id)query
viewId:(NSString *)viewId rootView:(RCTRootView *)rootView {
        self = [self initWithBridge:bridge componentName:componentName path:path
query:query viewId:viewId];
        // 将传入的已经渲染过的rootView保存起来
        self.rootView = rootView;
        return self;
    }

    - (void)viewDidLoad {
        [super viewDidLoad];
        // ...
        if (!self.rootView) {
            // 非预渲染逻辑
            [self creatRootView];
        } else {
            // 预渲染逻辑
            // 同步参数到JavaScript端
            [self.rootView setAppProperties:[self createAppProps]];
        }

        [self.view addSubview:self.rootView];
        // ...
    }

    @end
```

使用视图预加载也需要注意一些问题，比如iOS在创建RCTRootView时，会直接触发React的相关生命周期，也就是会导致componentDidMount在不可见时就触发，从而会导致一些统计方面的问题；另外，当执行[RCTRootView setAppProperties:props]时，也会重新触发JavaScript端的componentDidMount。

3. Android

对于 Android React Native 视图的创建流程，比较关键的部分就是 ReactNativeDelegate 类中的 loadApp 方法。在其中创建一个 ReactRootView 实例并调用它的 startReactApplication 方法完成 JavaScript 端视图的渲染，然后调用 Activity 的 setContentView 方法将渲染好的视图填充至 Activity 中，关键代码如下。

```java
public class ReactActivityDelegate {
    ......
    protected void loadApp(String appKey) {
        if (mReactRootView != null) {
            throw new IllegalStateException("Cannot loadApp while app is already running.");
        }
        // 创建 ReactRootView 实例
        mReactRootView = createRootView();
        // 触发 JavaScript 端生命周期渲染视图
        mReactRootView.startReactApplication(
            getReactNativeHost().getReactInstanceManager(), // ReactInstanceManager 实例
            appKey, // JavaScript 端通过 AppRegistry 注册的 MainComponentName
            getLaunchOptions()); // 传入的初始 props 数据
        // 将 ReactRootView 实例设置为当前 Activity 的内容视图
        getPlainActivity().setContentView(mReactRootView);
    }
}
```

为了实现视图预加载，需要缓存和管理由 JavaScript 端视图渲染的 ReactRootView，在 Activity 创建时从缓存池中按照一定的规则找到缓存的 ReactRootView 实例，并将它作为当前 Activity 的内容视图。为此，下面设计了一个视图加载类 PageLoader，它负责预缓存要显示的视图，并且提供匹配视图的方法。如果未匹配到缓存池中的视图，则新建一个视图；如果匹配到了缓存池中的视图，则将视图提供给 Activity 进行展示。具体关键代码如下。

```java
public class PageLoader {
    private static PageLoader sPageloader;
    private Hashtable<String, ReactRootView> mCachedViewPool = new Hashtable<>();
    private ReactApplication mReactApplication;

    private PageLoader() {}

    public static PageLoader getInstance() {
```

```java
        if (sPageloader == null) {
            synchronized (PageLoader.class) {
                if (sPageloader == null) {
                    sPageloader = new PageLoader();
                }
            }
        }
        return sPageloader;
    }

    public void init(ReactApplication application) {
        mReactApplication = application;
    }

    /**
     * 缓存一个待显示的视图
     *
     * @param context             上下文对象
     * @param componentName JavaScript 端 AppRegistry 注册的 mainComponentName
     */
    public void cache(Context context, String componentName) {
        if (contains(componentName)) {
            Log.e("PageLoader", componentName + "已在缓存池中，跳过本次缓存步骤");
            return;
        }
        ReactRootView reactRootView = createReactRootView(context, componentName);
        mCachedViewPool.put(componentName, reactRootView);
    }

    /**
     * 获取一个要显示的 ReactRootView 实例
     * 如果缓存池中存在匹配的 ReactRootView 实例则优先使用，如果不存在则新建一个 ReactRootView
     实例并返回
     *
     * @param context             上下文对象
     * @param componentName JavaScript 端 AppRegistry 注册的 mainComponentName
     * @return 要显示的 ReactRootView 实例
     */
    public ReactRootView getReactRootView(Context context, String componentName) {
        if (!contains(componentName)) {
```

```java
                    Log.e("PageLoader", "getReactRootView: " + componentName + "未缓存,
开始创建新的页面");
            return createReactRootView(context, componentName);
        }
        return mCachedViewPool.get(componentName);
    }

    /**
     * 在 Activity 销毁时调用,传入即将销毁的 ReactRootView 实例
     * 若当前视图需要循环使用,则在被销毁后重新创建 ReactRootView 实例,并放置在缓存池中
     * 若不需要则在缓存池中将它移除
     *
     * @param view 即将销毁的 ReactRootView 实例
     */
    public void recycleOrRelease(ReactRootView view, String componentName,
boolean recycle) {
        if (recycle) {
            mCachedViewPool.put(
                componentName,
                createReactRootView(
                    view.getContext(),
                    componentName
                )
            );
        } else {
            mCachedViewPool.remove(componentName);
        }

    }

    /**
     * 判断缓存池中是否已经具有对应的视图
     * 注意:暂不支持缓存可复用的页面,如果有需要,请修改此方法的实现
     *
     * @param componentName JavaScript 中 AppRegistry 注册的 mainComponentName
     * @return 是否已经具有对应的视图的结果
     */
    public boolean contains(String componentName) {
        return mCachedViewPool.containsKey(componentName) && mCachedViewPool.
get(componentName) != null;
```

```
    }

    /**
     * 创建 ReactRootView 实例
     *
     * @param context          上下文对象
     * @param componentName JavaScript端 AppRegistry 注册的 mainComponentName
     * @return 创建的 ReactRootView 实例
     */
    private ReactRootView createReactRootView(Context context, String componentName) {
        if (mReactApplication == null) {
            Log.e("PageLoader", "createReactRootView: 尚未初始化！");
        }
        ReactRootView reactRootView = new ReactRootView(context);
        reactRootView.startReactApplication(
            mReactApplication.getReactNativeHost().getReactInstanceManager(),
            componentName
        );
        return reactRootView;
    }

}
```

接下来需要新建一个代理类，继承ReactNativeDelegate并改写loadApp方法，可以使用刚刚编写的ReactRootViewLoader来加载视图，具体的代码如下。

```
public class PreloadDelegate extends ReactActivityDelegate {

    private ReactRootView mReactRootView;

    public PreloadDelegate(ReactActivity activity, @Nullable String mainComponentName) {
        super(activity, mainComponentName);
    }

    @Override
    protected void loadApp(String appKey) {
        // 这里通过 PageLoader 类来获取 ReactRootView
        mReactRootView = PageLoader.getInstance().getReactRootView(getPlainActivity(), getMainComponentName());
```

```
            getPlainActivity().setContentView(mReactRootView);
    }

    @Override
    protected void onDestroy() {
        PageLoader.getInstance().recycleOrRelease(mReactRootView,
getMainComponentName(), true);
        if (mReactRootView != null) {
            mReactRootView.unmountReactApplication();
            mReactRootView = null;
        }
        if (getReactNativeHost().hasInstance()) {
            getReactNativeHost().getReactInstanceManager().onHostDestroy(getPla
inActivity());
        }
    }
}
```

最后改写 ReactActivity 的 createReactNativeDelegate 方法,返回刚刚新建的代理类的实例即可,具体的代码实现如下。

```
public class RNPreloadExampleActivity extends ReactActivity {
    ......
    @Override
    protected ReactActivityDelegate createReactActivityDelegate() {
        return new PreloadDelegate(this, getMainComponentName());
    }
}
```

完成了以上代码后,可以在 JavaScript 端绘制一个较为复杂的 UI 布局,然后使用视图预加载和普通方式进行页面跳转,从而进行验证。可以看出,没有使用预加载的页面跳转会导致短暂的白屏,而视图预加载可以完美地解决这样的问题,达到"秒开"的效果。

与 iOS 不同的是,Android 更新 props 执行 ReactRootView.setAppProperties 时,并不会触发 JavaScript 端的 componentDidMount,仅触发生命周期中的 componentWillReciveProps 来刷新视图。

11.3 长列表优化

长列表其实是 React Native 经常被诟病的一个特性,作为一个诞生数年的框架,对于这个不算生僻的业务场景,React Native 至今还没有提供一个完美的解决方案。虽然目前 React Native 已经在 ScrollView

的基础上提供了 FlatList、SectionList 等做过优化的组件，基本覆盖了大部分业务场景，但对于某些极端场景（低性能手机、高速滚动），也只能做降级处理。本节先具体分析目前 React Native 对长列表所做的优化策略，之后会结合原生能力提供一些其他的思路。虽然这些方法不能覆盖所有应用场景的问题，但在特定情况下也能达到令人较为满意的优化结果。

11.3.1　FlatList、SectionList 和 VirtualizedList

React Native 中最基础、最简单粗暴的列表组件是 ScrollView，该组件会把所有的子元素一次性渲染出来，所以在需要使用长列表的场景中基本不会考虑该组件，否则在初始化时就会出现一段时间的白屏现象。FlatList 和 SectionList 则封装了底层的 VirtualizedList 组件，实现了一个惰性加载列表。所谓的惰性加载列表指组件本身维护一个有限行数的实际渲染窗口（通常会比屏幕略大一些），并将渲染窗口外的元素全部用适当定长的空白控件代替，这样从一定程度上减少了内存的消耗。同时，渲染窗口也会监听列表的滚动事件，根据滚动的距离获取对应数据重新渲染列表元素，并根据元素距离可视区域的远近设定渲染的优先级，以便在滚动过程中减少出现空白区域的可能。就 VirtualizedList 组件而言，为了减少内存占用以保持滑动的流畅性，渲染窗口外的内容会进行异步绘制，也就是说如果用户滑动列表的速度超过渲染的速度，列表就会出现短暂的白屏现象。

下面我们将具体解析 VirtualizedList 组件，从实现层面分析及借鉴其优化的实现方式。

首先需要注意的是 VirtualizedList 继承于 React 的 PureComponent，也就是说组件的 props 是浅比较。在 renderItem 中依赖的数据如果是引用类型（例如 Object 或 Array）且需要更新，则必须修改其引用地址，创建新的 Object 或 Array，否则很可能导致页面并不刷新。

其次分析组件的渲染逻辑，上文也说了 VirtualizedList 是一个惰性加载的列表，那在 render 执行时，会有一段较为复杂的策略，例如：

```
render() {
  ......
  if (itemCount > 0) {
  _usedIndexForKey = false;
  _keylessItemComponentName = '';
  const spacerKey = !horizontal ? 'height' : 'width';
  const lastInitialIndex = this.props.initialScrollIndex // initialScrollIndex,
设置初始化渲染开始的索引
    ? -1
    : this.props.initialNumToRender - 1; // initialNumToRender, 初始化渲染的元素数量
  const {first, last} = this.state;
  // 渲染初始化时首屏需要的组件数
  this._pushCells(
    cells,
```

```
      stickyHeaderIndices,
      stickyIndicesFromProps,
      0,
      lastInitialIndex,
      inversionStyle,
    );
    const firstAfterInitial = Math.max(lastInitialIndex + 1, first);
    // 填充顶部距离
    if (!isVirtualizationDisabled && first > lastInitialIndex + 1) {
      let insertedStickySpacer = false;
      // 处理存在悬浮组件的情况
      if (stickyIndicesFromProps.size > 0) {
        const stickyOffset = ListHeaderComponent ? 1 : 0;
        for (let ii = firstAfterInitial - 1; ii > lastInitialIndex; ii--) {
          ......
        }
      }
      if (!insertedStickySpacer) {
        const initBlock = this._getFrameMetricsApprox(lastInitialIndex);
        const firstSpace =
          this._getFrameMetricsApprox(first).offset -
          (initBlock.offset + initBlock.length);
        cells.push(
          <View key="$lead_spacer" style={{[spacerKey]: firstSpace}} />,
        );
      }
    }
    // 渲染当前屏幕可见的组件数
    this._pushCells(
      cells,
      stickyHeaderIndices,
      stickyIndicesFromProps,
      firstAfterInitial,
      last,
      inversionStyle,
    );
    ......
    // 填充底部距离
    if (!isVirtualizationDisabled && last < itemCount - 1) {
      const lastFrame = this._getFrameMetricsApprox(last);
```

```
    const end = this.props.getItemLayout
      ? itemCount - 1
      : Math.min(itemCount - 1, this._highestMeasuredFrameIndex);
    const endFrame = this._getFrameMetricsApprox(end);
    const tailSpacerLength =
      endFrame.offset +
      endFrame.length -
      (lastFrame.offset + lastFrame.length);
    cells.push(
      <View key="$tail_spacer" style={{[spacerKey]: tailSpacerLength}} />,
    );
  }
}
  ......
}
```

最后需要注意的就是组件的滚动事件，绑定的 onScroll 会根据当前滚动到的位置及整个组件的长（宽）度来计算应该渲染的具体行数，具体代码如下。

```
_onScroll = (e: Object) => {
    this._nestedChildLists.forEach(childList => {
      childList.ref && childList.ref._onScroll(e);
    });
    if (this.props.onScroll) {
      this.props.onScroll(e);
    }
    const timestamp = e.timeStamp;
    let visibleLength = this._selectLength(e.nativeEvent.layoutMeasurement);
// 可见区域长度
    let contentLength = this._selectLength(e.nativeEvent.contentSize); // 内容长度
    let offset = this._selectOffset(e.nativeEvent.contentOffset); // 内容位置偏移量
    let dOffset = offset - this._scrollMetrics.offset;
    ......
    const dt = this._scrollMetrics.timestamp
      ? Math.max(1, timestamp - this._scrollMetrics.timestamp)
      : 1;
    const velocity = dOffset / dt;
    ......
    // 滚动相关的测量数据
    this._scrollMetrics = {
```

```
            contentLength,
            dt,
            dOffset,
            offset,
            timestamp,
            velocity,
            visibleLength,
        };
        this._updateViewableItems(this.props.data);
        if (!this.props) {
          return;
        }
        this._maybeCallOnEndReached();
        if (velocity !== 0) {
          this._fillRateHelper.activate();
        }
        this._computeBlankness(); // 计算空白区域
        // 根据当前位置重新计算并更新state中的first、last，触发render
        this._scheduleCellsToRenderUpdate();
    };
```

11.3.2　原生视图的复用

在原生开发中，长列表并不属于一个性能的难点，iOS中的UITableView、Android中的RecyclerView都很好地解决了这一点。在这些组件中，列表渲染主要分为视图布局搭建和填充数据两个步骤。前者负责制定UI组件结构；后者负责填充组件内容（例如文本内容和图片地址等），并且在实现滚动时可以复用每一行的视图布局，通过不断地改变UI组件内的填充数据达到滑动的效果。这样就省去了每次UI组件重新构建的过程，从而得到流畅、顺滑的视觉效果。

React Native是否可以利用原生组件的这种复用方式提升自己的性能呢？为了验证这一设想，可以开发一个自定义的原生UI组件，封装UITableView或RecyclerView组件，再尝试利用这两个组件复用每一行的组件。

1. iOS

首先介绍一下UITableView和UITableViewCell这两个iOS的原生列表中的类。UITableView通过注册UITableViewCell的方式对Cell元素进行复用，根据屏幕的高度和预先设置好的Cell高度进行实际数量的创建。比如当屏幕只能展示完整的2个Cell时，那么系统会在屏幕外的上下两侧预留出1个Cell，也就是说实际一共会创建4个Cell，在页面滚动时就使用这4个Cell进行复用。

一般对于iOS的长列表开发，多个列表样式会对应多个Cell，每个Cell会使用不同的key以保证视图和页面的绑定关系；并且当Cell能够定高时，复用的效果最好。

如果想在React Native环境中使用这种复用机制，大致有以下两种思路。

（1）沿用原生的理念，根据不同的使用场景定制不同的Cell，将UITableView封装成React Native UI组件，并暴露属性供JavaScript端选择需要展示的Cell类型，以及由JavaScript端提供数据源驱动原生页面刷新。这个方案的本质是由原生去绘制和渲染UI组件，组件性能跟原生开发没有太大差异。但缺点也很明显，JavaScript端无法自定义组件样式，一旦产生变更，就需要原生端重新开发。

（2）iOS仅提供一个基础Cell，同样将UITableView封装成React Native UI组件，但是Cell通过React Native端AppRegistry注册的方式生成。iOS则通过注册的moduleName在Cell中创建RCTRoottView，并且在数据更新时主动刷新每个React Native Cell。这个方案明显要更为灵活，且iOS只需一次开发，若需要多个UI样式，只需在JavaScript端注册多个组件。不过这种方案也不是没有缺点，在这种列表中每个Cell都是RCTRootView，拥有完整的Touch事件监听流程，与列表自身的RCTRootView会产生重合，导致事件的透传或丢失。

我们可以将第二种方案归纳成以下流程，如图11.6所示。

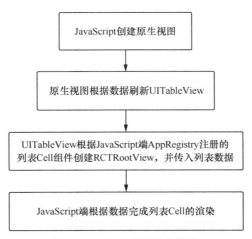

图11.6 iOS长列表优化流程

前面的章节已经介绍过自定义原生UI组件的方式，这里也采用同样的方法对iOS的UITableView进行封装。

```
// ListViewManager.m

// 声明组件名称
RCT_EXPORT_MODULE(LargeList)

// 声明组件属性
```

```
RCT_CUSTOM_VIEW_PROPERTY(data, NSArray, LargeListView) {
    [view tableViewReloadDataWith:json];
}

RCT_EXPORT_VIEW_PROPERTY(itemModuleName, NSString)

RCT_EXPORT_VIEW_PROPERTY(itemHeight, CGFloat)

- (UIView *)view {
    LargeListView *view = [[LargeListView alloc] initWithBridge:self.bridge];
    return view;
}
```

其中,RCT_CUSTOM_VIEW_PROPERTY 可以监听 JavaScript 端属性的变化,用于处理数据源切换后的重新渲染。LargeListView 则是实际分装 UITableView 的 View 对象,其中创建的方式与原生无差别,大致代码如下。

```
// LargeListView.m

……

@implementation LargeListView

- (instancetype)initWithBridge:(RCTBridge *)bridge
{
    self = [super initWithFrame:CGRectZero];
    if (self) {
        _bridge = bridge;
        [self createSubViews];
    }
    return self;
}

// 创建UITableView视图
- (void)createSubViews {
    _tableView = [[UITableView alloc] initWithFrame:self.bounds style:UITableViewStylePlain];
    _tableView.delegate = self;
    _tableView.dataSource = self;
    _tableView.separatorStyle = UITableViewCellSeparatorStyleNone;
```

```objc
    [_tableView registerClass:[ListTableViewCell class] forCellReuseIdentifier:
KCustomTableViewCellID];

    [self addSubview:_tableView];
}
// 设置Cell
- (UITableViewCell *)tableView:(UITableView *)tableView cellForRowAtIndexPath:
(NSIndexPath *)indexPath {
    ListTableViewCell *cell = [tableView dequeueReusableCellWithIdentifier:
KCustomTableViewCellID];
    cell.selectionStyle = UITableViewCellSelectionStyleNone;
    NSDictionary *content = _dataSource[indexPath.row];
    // 渲染实际的Cell视图
    [cell setupBridge:_bridge withModuleName:_itemModuleName withData:content];
    return cell;
}

……

// 数据更新
#pragma mark -refresh
- (void)tableViewReloadDataWith:(NSArray *)dataSource {
    _dataSource = dataSource;
    [_tableView reloadData];
}
// 在LargeListView中的视图刷新时，需要修改_tableView的frame，
// 因为在初始化时，并不会真正获取到JavaScript端设置的style
- (void)layoutSubviews {
    [super layoutSubviews];
    _tableView.frame = self.bounds;
}

@end
```

实际的Cell代码如下。

```objc
//
// ListTableViewCell.m
……

@implementation ListTableViewCell
```

```objc
    // 这里的参数是viewManager中传递给LargeList的bridge对象
    // JavaScript端定义的name组件和数据源
- (void)setupBridge:(RCTBridge *)bridge withModuleName:(NSString *)moduleName
withData:(NSDictionary *)data {
    BOOL isLegal = [data isKindOfClass:[NSDictionary class]] && [moduleName
isKindOfClass:[NSString class]];
    if (!isLegal) return;
    [_rootView removeFromSuperview];
    _rootView = nil;
    _moduleName = moduleName;
    // 重新渲染组件并加载到Cell上
    _rootView = [[RCTRootView alloc] initWithBridge:bridge moduleName:moduleName
initialProperties:data];
    [self.contentView addSubview:_rootView];

}
......
@end
```

在使用UITableView时，通常需要在创建前返回Cell的高度，在纯粹原生场景中可以先给出Cell的预估高度，然后在实际创建列表后再对高度值进行更新。但在当前使用RCTRootView的写法下，我们无法在未创建组件时获取到Cell组件的高度，所以预留了一个itemHeight的参数，让JavaScript获取到Cell组件高度后再创建整个列表。

2. Android

Android开发中常使用RecyclerView实现长列表的需求。它采用了两级缓存的方案，能够将滑出屏幕的条目缓存和复用，所以在计算大量数据的情况下，列表也能够保持很好的性能。RecyclerView的缓存方案具有以下特点。

（1）对于刚刚滑出屏幕的条目，此方案会将其从父布局回收，然后原样放在第一级缓存池中。此时如果用户想再将其滑入屏幕中，即可直接从第一级缓存池中取出该条目，添加至父布局快速显示出来。

（2）如果用户一直滑动，回收的条目将第一级缓存池填充满了，此方案则会将离屏幕最远的条目中的UI状态（例如文本内容、图片地址等）清除，只保留UI布局放置在第二级缓存池中。此时如果即将滑入屏幕的条目需要的UI布局与第二级缓存池中的条目布局一致的话，优先复用第二级缓存池中的条目，为其重新填充新的数据，添加至父布局显示出来。

（3）如果用户一直滑动，回收的条目将第二级缓存池也填充满了，此方案则会将第二级缓存池中最近最少使用的UI布局的条目销毁，节约内存。此时如果即将滑入屏幕的条目需要的UI布局在第二级缓存池的条目中没有匹配的话，才会重新创建条目对象，初始化布局，再填充数据展示出来，与没有缓存方案的情况一致。

为了方便开发者快速实现上述的两级缓存方案，Android为开发者提供了RecyclerView.Adapter这个抽象类，开发者只需要实现3个关键的抽象方法即可。

onCreateViewHolder：在此方法内实例化条目对象，完成布局的初始化，并将条目对象添加进ViewHolder类中。

onBindViewHolder：调用ViewHolder类的数据绑定方法完成对条目对象UI数据的填充。

getItemCount：返回列表数据的整体长度。

前几章讲解了JavaScript端定义的每个Component组件，经过AppRegistry注册后，这些组件在Android端都可以创建一个ReactRootView实例，调用ReactRootView.startReactApplication方法，将AppRegistry注册的MainModuleName传入后，即可触发React生命周期，完成JavaScript定义视图的渲染。

综合RecyclerView的二级缓存方案以及Android端渲染JavaScript定义视图的流程，我们可以定义一个RecyclerView.Adapter的实现类，在onCreateViewHolder方法中创建ReactRootView实例，在onBindViewHolder方法中调用ReactRootView.startReactApplication方法传入要渲染的JavaScript端视图以及要传入的数据即可，具体的实现如下。

```java
public class ListAdapter extends RecyclerView.Adapter<ListAdapter.ListItemViewHolder> {

    public ListAdapter(ReactInstanceManager reactInstanceManager, Context context) {
        mReactInstanceManager = reactInstanceManager;
        mContext = context;
    }

    @NonNull
    @Override
    public ListItemViewHolder onCreateViewHolder(@NonNull ViewGroup parent, int viewType) {
        // 创建 ReactRootView 对象
        ReactRootView rootView = new ReactRootView(mContext);
        // 将其放入 ViewHolder 类中
        return new ListItemViewHolder(rootView);
    }

    @Override
    public void onBindViewHolder(@NonNull ListItemViewHolder holder, int position) {
        // 构造即将为布局绑定的数据
        Bundle props = new Bundle();
```

```java
            props.putString("text", data.get(position).text);
            // 调用 ViewHolder 类的绑定方法完成数据的绑定
            holder.bind(props);
        }

        private void startApp(ReactRootView rootView, Bundle props) {
            if (rootView == null) {
                rootView = new ReactRootView(mContext); // 内部的布局和渲染交给 React Native 完成
            }
            // 此处指定了高度等布局信息,可以保证列表在滚动时的性能更好
            rootView.setLayoutParams(new RecyclerView.LayoutParams(ViewGroup.LayoutParams.MATCH_PARENT, getItemHeight()));
            rootView.startReactApplication(mReactInstanceManager, getItemModuleName(), props);
        }

        @Override
        public int getItemCount() {
            return data == null ? 0 : data.size();
        }

        // ......

        class ListItemViewHolder extends RecyclerView.ViewHolder {

            private ReactRootView mItemView;

            public ListItemViewHolder(@NonNull ReactRootView itemView) {
                super(itemView);
                mItemView = itemView;
            }

            public void bind(Bundle props) {
                if (mItemView.getReactInstanceManager() == null) {
                    // 如果此条目是新创建的,未绑定过数据,调用 startReactApplication 方法
                    // 完成页面渲染
                    startApp(mItemView, props);
                } else {
                    // 如果此条目是被回收复用的,则调用 setAppProperties 方法完成绑定数据的
```

更新，触发页面渲染

```
                mItemView.setAppProperties(props);
        }
    }
}
```

3. JavaScript

上述原生组件都需要通过实例化RCTRootView/ReactRootView的方式创建列表条目，所以在JavaScript端需要通过AppRegistry. registerComponent/registerSection的方式注册列表条目视图，例如：

```
AppRegistry.registerComponent('11_3_items',() => Example11_3_items);

// Example11_3_items
import React, { Component } from 'react';

import {
  Text,
  View,
} from 'react-native'

export default class App extends Component{
  constructor(props){
    super(props)
  }

  render() {
    return (
    <View style={{ flex: 1, justifyContent: 'center', alignItems: 'center' }}>
      <Text>{this.props.text}</Text>
    </View>
    );
  }
}

// 列表组件使用方式
import React, { Component } from 'react';
import {
```

```jsx
    SafeAreaView,
    View,
    requireNativeComponent,
    Dimensions
} from 'react-native'

const screenSize = Dimensions.get('window');
const DEVICE_WIDTH = screenSize.width;
const DEVICE_HEIGHT = screenSize.height;
// 生成1000条数据
let count = 0;
const list = Array.from(new Array(1000), () => {
  count++;
  return {text: '这是由JavaScript绘制的内容' + count}
});
const NativeList = requireNativeComponent("LargeList");

export default class App extends Component{
  constructor(props){
    super(props)
  }

  render() {
    return (
      <SafeAreaView style={{flex: 1}}>
        <View style={{ flex: 1, justifyContent: 'center', alignItems: 'center'}}>
          <NativeList
            style={{ height: DEVICE_HEIGHT, width: DEVICE_WIDTH }}
            itemHeight={50}
            itemModuleName={'11_3_items'}
            data={list}
          />
        </View>
      </SafeAreaView>
    );
  }
}
```

这套实现方案基本摆脱了数量级对于列表的影响，在千级别的图文列表中也有很好的表现，尤其是在Android的某些低端设备中，基本不会出现FlatList和SectionList中由于快速滑动而造成的白屏现象。不过由于列表条目均是独立的 RCTRootView/ReactRootView，在初始化时耗费的时间会多一些。

11.4 Tab 切换

对于 Tab 切换这个场景，如果简单处理并不存在太大的性能瓶颈，只需要销毁当前 Tab 内容并新建下一个需要呈现的 Tab 内容。但通常为了提升体验，我们经常会把这个场景设计成具有手势切换动画、缓存单项 Tab 内容等其他附加功能。这个场景也相当通用，在原生 UI 组件中都可以找到直接实现的类，例如 iOS 的 UITableView、Android 的 RecyclerView，都提供了很好的机制优化切换这种场景。

React Native 最初也提供了 ViewPagerAndroid 这样封装原生 UI 的组件（目前已被移除，需要使用的话需单独引入 react-native-community/react-native-viewpager），主要包括了一个允许子视图左右翻页的容器。每一个 ViewPagerAndroid 的子容器会被视作一个单独的页面，并且会被拉伸填满 ViewPagerAndroid，且子视图都必须是纯视图，而不能是自定义的复合容器，例如：

```
render() {
  return (
    <ViewPagerAndroid
      style={styles.container}
      initialPage={0}
    >
      <View style={styles.tab} key="1">
        <Text>第一页</Text>
      </View>
      <View style={styles.tab} key="2">
        <Text>第二页</Text>
      </View>
    </ViewPagerAndroid>
  );
}

......

const styles = {
  container: {
    flex: 1
  },
  tab: {
    justifyContent: 'center',
```

```
        alignItems: 'center',
    }
}
```

11.5 本章小结

在硬件资源一定的情况下,优化更多的是从策略方面做调整。在不影响用户感知的情况下,可以将一些后置操作提前运行,或者将高消耗的操作结果缓存复用,从而为用户提供效果操作及视觉效果。以UI组件为例,计算位置、绘制渲染无疑是最消耗性能的过程,如果能将绘制好的UI组件复用,或对超出屏幕范畴的UI降级处理,那无疑能减少资源的消耗,避免出现卡顿的现象。这些策略有的是平台组件本身就具备的,有的则依靠框架实现,有的就需要开发者有意识地调整。React Native的重要功能是解决跨平台问题,对于特定场景也不会进行过于个性化的优化,否则可能会破坏整体的开发形式,所以也就需要开发者了解其内部的机制再进行针对性的调优。

第12章　React Native中的"微前端"

2019年，前端生态中最火的两个概念莫过于Serverless（无服务器）和微前端了，D2论坛也分别开设了两个专场来分享这两个概念及目前厂商对其的实践。两者目前都没有明确的官方定义，Serverless并不是指真的不需要服务器，而是指开发者并不需要关心这些基础架构的存在，只需要提供代码或函数即可。微前端则是为了能将一个巨大的前端工程拆分成一个个小型工程而设计的架构模式，并且这些小工程具备独立的开发和运行能力，整个系统依赖于这些小工程的协同与合作，且目前绝大部分微前端的方案都是基于浏览器环境下的Web系统，尤其是前后端分离的单页面应用。

随着智能手机的普及、移动互联网的发展，开发一个移动端App已经不是什么技术难点，并且研发成本也越来越低。但如何让用户安装你的App，反而变成了一件成本巨大且极其困难的事情。没有人期望自己手机中的App越来越多，安装了几十个App却在临使用时找不到。所以，一些用户量极大的应用逐渐发展成了入口级、平台级的App。其中包含的业务和代码在量级上也都有明显的增加，这也对应用的工程和架构设计提出了新的要求。

那么，有可能在App中实现类似于微前端概念的方案吗，比如将一个App拆解成以业务为划分的小工程，且具备独立开发和部署的能力？本章就来介绍一下当前微前端的一些重要概念，以及在React Native中人们对于微前端方案的一些尝试。

12.1 什么是微前端

从概念上讲，微前端的概念主要来自于Microservices（微服务），是把应用设计成一系列松耦合的细粒度服务，并通过轻量级的通信协议组织起来；而且这些服务都能够独立部署、独立扩展，每个服务都具有稳固的模块边界，甚至允许使用不同的编程语言编写不同服务，也可以由不同的团队管理。

日益复杂的前端工程也面临着系统过于庞大、繁多的问题，特别是在toB业务中存在大量管理后台系统的场景。如果这些业务都放在同一个项目中开发和维护，不仅项目代码量巨大，调试、编译及打包成本也会逐步增加，甚至达到不可控的阶段，上线会变成一件极具风险的事情，因为你不知道你所做的修改是否会造成其他页面的问题。而如果每个业务都是一套独立的系统，则会导致代码的复用率偏低，一旦系统数量上升，不仅用户觉得入口繁多难以使用，整体维护的成本也会上升。

那在这种情况下，微前端方案会如何去设计这个整体架构呢？我们可以先参考图12.1所示的架构。

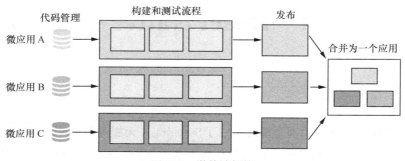

图 12.1　微前端架构

理想状态下，可以将项目分解成一个个独立的前端子应用，拥有自己的代码控制权限，以及发布、测试和上线流程，最后完成发布之后的这些微前端项目可以合并成一个整体的系统，这里暂时称之为主应用（或者叫宿主）。

从设计方案和期望来看，这种架构基本上解决了我们所面临的问题，那最关键的问题就出现了，我们如何来实现这样的设计方案？

整体来讲，目前业界有以下几种实现方案。

（1）npm：子应用以npm包的形式发布，自身的开发阶段则依赖于主应用中的工程管理。

（2）iframe：这种方案过于简单粗暴，每次跳转都会重新加载，主/子应用间很难复用和交互，且路由地址管理困难。

（3）使用HTTP服务器的路由来重定向多个应用：依赖于Nginx等服务器软件，对于特定路径加载特定的HTML入口文件，作为独立的应用。

（4）主应用通用路由：主应用依赖特定框架可以注册子应用，不强制统一技术栈，目前比较成熟的框架有single-spa、qiankun等。

（5）主应用特定路由：主应用依赖特定框架注册子应用，强制统一技术栈，这种方案实现起来会比通用路由简单，不用考虑不同技术栈之前的兼容问题。

后3种方式也被称为路由分发式微前端，即可以通过路由将不同的业务分发到不同的、独立的前端应用，只不过实现手段不一样，效果也有差异。下面以Nginx和single-spa两种方式讲述一下大致的实现过程。

1. Nginx分发路由

在单个应用中，路由通常是由框架内部实现的，例如React.js的react-router-dom，Vue.js的vue-router，通过监听URL的变化切换对应的组件。在这种设计中，整体的路由交给了Nginx控制，利用

location匹配特定路径，分发到特定的子应用，从而加载子应用所需的入口HTML文件和JavaScript资源等。例如：

```
http {
  server {
    listen       80;
    server_name  www.yourwebsite.com;
    location /api/ {
      proxy_pass http://127.0.0.1:8000/api;   // 数据接口
    }
    location /web/app1/ {
      index index.html;
      root /data/www/app1/;    // 应用A的入口文件路径
    }
    location /web/app2/ {
      index index.html;
      root /data/www/app2/;  // 应用B的入口文件路径
    }
    ……
  }
}
```

这种方式实现简单，成本也低，还不用关心子应用实现的技术栈是否相同，但实际效果上会有不少缺陷。

（1）跨应用之间的页面跳转只能依赖window.location.href，且依然会有页面整体加载的情况。

（2）各应用之间通信和同享数据较为困难，基本只能依赖于全局对象window。

（3）对于类似于导航、侧边栏之类的全局功能需要各应用自己实现，没有办法达到复用的效果。

2. single-spa

single-spa是目前较为成熟的微前端实现框架之一，它能够兼容不同的前端框架（例如React、Vue、Angular等），将它们当作子应用注册，从而使用统一的路由、状态来进行管理。各个子应用之间也不会互相影响，并且可以共享主应用的状态，进行通信。整体系统的架构如图12.2所示。

主应用就是一个调度中心，主要提供的功能如下。

（1）维护子应用的注册表。

（2）通过路由来控制子应用的各个生命周期。

（3）加载子应用的JavaScript Bundle文件，为子应用提供挂载点。

（4）提供全局变量、通信方式等。

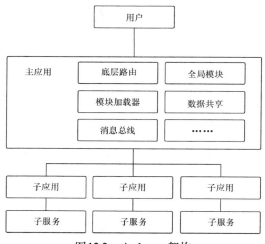

图12.2　single-spa架构

由于主应用中需要动态引入外部JavaScript文件，并且又不影响在开发时使用import、require方法，所有项目需要依赖一个systemjs的工具库，并且将子项目最终打包成UMD格式。在主应用中，需要这么一张注册表：

```
<script type="systemjs-importmap">
  {
    "imports": {
      "app1": "http://localhost:8081/app.js",  // 子应用文件
      "app2": "http://localhost:8082/app.js",  // 子应用文件
      // 主应用依赖
      "single-spa": "https://cdnjs.cloudflare.com/ajax/libs/single-spa/4.3.7/system/single-spa.min.js",
      "vue": "https://cdn.jsdelivr.net/npm/vue@2.6.10/dist/vue.js",
      "Vue": "https://cdn.jsdelivr.net/npm/vue@2.6.10/dist/vue.js",
      "vue-router": "https://cdn.jsdelivr.net/npm/vue-router@3.0.7/dist/vue-router.min.js"
    }
  }
</script>
```

然后注册子应用：

```
<script>
  (function() {
    Promise.all([System.import('single-spa'), System.import('vue'), System.import('vue-router')]).then(function (modules) {
```

```
      var singleSpa = modules[0];
      var vue = modules[1];
      var vueRouter = modules[2];
      vue.use(vueRouter)
      // 注册子应用
      singleSpa.registerApplication(
        'app1',
        () => System.import('app1'),
        location => location.pathname === '/' && location.hash.startsWith('#/app1')
      )

      singleSpa.registerApplication(
        'app2',
        () => System.import('app2'),
        location => location.pathname === '/' && location.hash.startsWith('#/app2')
      )

      /**
       *  检测模块错误，方便查看报错信息
       */
      singleSpa.addErrorHandler(err => {
        console.log(err);
        console.log(err.appOrParcelName);
        console.log(singleSpa.getAppStatus(err.appOrParcelName));
      });
    })
  })()
</script>
```

可以看出主应用会检测路由的变化决定引用哪个子应用的 JavaScript Bundle 文件，例如子应用 app1 的加载条件就是 location 满足条件：location.pathname==='/' && location.hash.startsWith('#/app1')。

在子组件中，需要使用 single-spa 对现有的应用进行一层包装，并且导出3个生命周期函数。下面以 Vue 项目为例：

```
import './set-public-path'
import Vue from 'vue'
import App from './App.vue'
import router from './router'
import store from './store'
import singleSpaVue from 'single-spa-vue'
```

```js
// 这里按照single-spa-vue的方式渲染
const vueLifecycles = singleSpaVue({
  Vue,
  appOptions: {
    el: '#app1',
    store,
    router,
    render: h => h(App)
  }
})

// 这里要输出必需的3个生命周期函数
export const bootstrap = vueLifecycles.bootstrap
export const mount = (props) => {
  // 这里可以获取主应用传递到子系统的变量或通过window._store获取主应用的store
  for (const key in props.store) {
    console.log('mount install store……', props)
    store.state[key] = props.store[key]
  }
  return vueLifecycles.mount(props)
}

export const unmount = vueLifecycles.unmount
```

其中的bootstrap、mount、unmount分别为启动、挂载和卸载，另外还有一个update生命周期函数，但不是必需的。

最后，需要设置子应用的打包机制，输出成符合UMD格式规范的文件，例如：

```js
// vue.config.js
/**
 * 在configureWebpack中，要将output的libraryTarget选项设为'umd'，以适应systemjs的加载方式
 */
configureWebpack: {
  resolve: {
    extensions: [ '.js', '.json', '.vue' ]
  },
  output: {
    libraryTarget: 'umd',
    filename: '[name].js'
  },
  externals: {
    // 'vue': 'Vue',
```

```
      // 'VueRouter': 'vue-router',
      // 'vuex': 'Vuex',
    }
  },
```

12.2 React Native "微前端" 探索

基于上一节对微前端的介绍，以及前期对混合存储、路由及多实例的一些方案探索，我们可以发现React Native环境其实具备了路由分发式微前端方案的条件；而且不需要考虑Nginx方式（条件不具备）和通用路由方案（没有必要），就可以直接采用主应用特定路由的思路。

我们可以将原生应用看成是微前端方案中的主应用，多个React Native的JavaScript Bundle文件就是其中的子应用，利用多实例和混合路由方案可以唤起任意属于不同Bundle文件的React Native视图，混合存储方案则能保证主、子应用中的数据可以共享。React Native微前端的大致架构如图12.3所示。

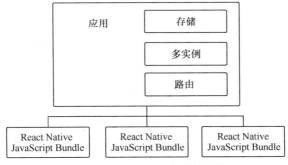

图12.3　React Native微前端架构

以iOS为示例，整体的实现可以大致分为以下几个步骤。

（1）在应用中设置多个React Native实例，类似于主应用中的注册表，并且由于Bridge天然就是隔离的，因此注册的各个子应用之间不会互相影响，例如：

```
// AppDelegate.m
……
- (BOOL)application:(UIApplication *)application didFinishLaunchingWithOptions:(NSDictionary *)launchOptions
{
    // 配置两个不同Bundle Name的React Native实例，相当于两个子应用
    MCRNBridgeConfig *config = [[MCRNBridgeConfig alloc] init];
    config.bundleName = @"bundle1";
    [[MCRNBridgeMgr shared] lazyCreateBridgeForConfig:config];
```

```
    MCRNBridgeConfig *config2 = [[MCRNBridgeConfig alloc] init];
    config2.bundleName = @"bundle2";
    [[MCRNBridgeMgr shared] lazyCreateBridgeForConfig:config2];
    ......
}
......
```

（2）使用混合路由加载页面，并且可以指定Bundle名称。

```
// iOS端直接调用
// 指定Bundle 名称
RCTBridge *bridge = [[MCRNBridgeMgr shared] bridgeForBundleName:@"bundle1"];
// 指定对应路由路径
MCRNRootView *rnView = [[MCRNRootView alloc] initWithBridge:bridge componentName:
@"app1" path:@"/partial" query:@"" viewId:@""];

// 在JavaScript端直接调用
import { history } from '@mcrn/bridge'
// bundle2为另一个
history.push('/about', {msg: 'msgFromMain'}, '', { navBarTextColor: 'black'},
'bundle2')
```

（3）React Native的事件发送机制天然支持主应用（App）、React Native子应用通信，也可以再进行一层封装，支持指定Bundle名称，以实现子>主>子方式的通信。

（4）采用混合存储方案共享数据，由于数据最终存放在原生应用中，因此任意子应用均可以使用。

```
// iOS
[MCRNStorage setItem:item forKey:key];
// JavaScript
MCRNStorage.getItem(key);
```

从开发流程上来看，完全可以将React Native子应用分别交给不同的团队，各自控制代码权限及进行开发、维护和上线。但如果出现子应用的原生依赖互相冲突的情况，还是需要人工协调解决，并不能做到完全隔离。

12.3 本章小结

微前端的核心目标是将Monolith（巨石）拆解成若干可以自治的松耦合微应用，这其实也表示了这种方案的适用场景。如果你的应用没有复杂到包含多条业务线、多个团队共同开发的情况，就不需要这种方案，否则只会增加自己的开发成本。